SpringerBriefs in Applied Sciences and Technology

Computational Intelligence

Series editor

Janusz Kacprzyk, Warsaw, Poland

About this Series

The series "Studies in Computational Intelligence" (SCI) publishes new developments and advances in the various areas of computational intelligence—quickly and with a high quality. The intent is to cover the theory, applications, and design methods of computational intelligence, as embedded in the fields of engineering, computer science, physics and life sciences, as well as the methodologies behind them. The series contains monographs, lecture notes and edited volumes in computational intelligence spanning the areas of neural networks, connectionist systems, genetic algorithms, evolutionary computation, artificial intelligence, cellular automata, self-organizing systems, soft computing, fuzzy systems, and hybrid intelligent systems. Of particular value to both the contributors and the readership are the short publication timeframe and the world-wide distribution, which enable both wide and rapid dissemination of research output.

More information about this series at http://www.springer.com/series/10618

Xiaolei Wang · Xiao-Zhi Gao
Kai Zenger

An Introduction
to Harmony Search
Optimization Method

 Springer

Xiaole Wang
Xiao-Zhi Gao
Kai Zenger
Department of Electrical Engineering
  and Automation
Aalto University
Espoo
Finland

ISSN 2191-530X   ISSN 2191-5318   (electronic)
ISBN 978-3-319-08355-1   ISBN 978-3-319-08356-8   (eBook)
DOI 10.1007/978-3-319-08356-8

Library of Congress Control Number: 2014943499

Springer Cham Heidelberg New York Dordrecht London

Printed on acid-free paper

Springer is part of Springer Science+Business Media (www.springer.com)

Contents

Chapter 1
Introduction

Abstract The development of modern science and technology promotes us to research more flexible and reliable problem-solving approaches. The nature-inspired computational (NIC) algorithms, alternatively called meta-heuristic algorithms, have attracted great attention in recent years. These nature-inspired algorithms, as the name implies, draw metaphorical inspirations from diverse natural sources, e.g., human thinking mechanism, physical principles, and social interaction. In the area of optimization, as an instance, their advantages including flexibility, randomness, robustness, mutually balanced in intensification and diversification, and intuitive guidelines determine their dominant position over the conventional optimization methods. This chapter gives a brief introduction to the NIC algorithms.

Keywords Nature-inspired computational algorithms · Meta-heuristic methods · Harmony search method · Hybrid nature-inspired computational methods

1.1 A Survey of Nature-Inspired Computational Methods

Inspired by natural phenomena and biological models, nature-inspired computational (NIC) methods have been devised to overcome the existing drawbacks of conventional numerical algorithms and witnessed enormous successes in various engineering optimization problem solving. Each of the NIC paradigms has its origins in biological systems from which the inspiration can be drawn. On the other word, these algorithms share two common characteristics in mimicking natural phenomena: inspiration drawn from nature and modeling of natural processes. For example, the collective behavior of unsophisticated agents interacting locally with their environment causes the coherent patterns of emergence [1]. These swarming, flocking, and herding phenomena have promoted a popular NIC method, known as swarm intelligence (SI), which was firstly coined by Beni and Wang [2, 3] in the 1980s in the context of cellular robotics [3]. Originally proposed by Metropolis in the early 1950s as a model of the crystallization process, the simulated annealing (SA) algorithm consists of first 'melting' the system being

© The Author(s) 2015
X. Wang et al., *An Introduction to Harmony Search Optimization Method*,
SpringerBriefs in Computational Intelligence,
DOI 10.1007/978-3-319-08356-8_1

Table 1.1 A taxonomy of NIC algorithms

Types	Algorithms
Human mechanism	Dendritic cell algorithm
	Clonal selection algorithm
	Negative selection algorithm
	Artificial immune networks
	Neural networks
Physical principle	Simulated annealing
	Harmony search
	Memetic algorithm
	Chaos optimization algorithm
	Cultural algorithm
	Charged system search
	Gravitational clustering algorithm
	Extremal optimization
	Water flow algorithm
Social interaction	Particle swarm optimization
	Ant colony optimization
	Fish swarm algorithm
	Artificial bee colony
	Cuckoo algorithm
	Collective animal behavior
Evolutionary computation	Evolution strategies
	Differential evolution
	Genetic algorithm
	Genetic programming
	Evolutionary programming
	Memetic algorithm
Other algorithms	Chemical reaction optimization
	Differential search
	Bacterial foraging algorithm
	Bionic optimization

optimized at a high temperature and then slowly lowering the temperature until the system 'freezes' and no further change occurs. In other words, the SA mimics the behavior of this dynamical system to achieve the thermal equilibrium at a given temperature [4]. Table 1.1 summarizes the widely utilized NIC algorithms in the classification of five groups: human mechanism, physical principles, social interaction, evolutionary computation, and other algorithms.

The NIC algorithms can be called nature-inspired meta-heuristic algorithms as well, and the optimal solution is achieved during the process of 'trial-and-error.' A meta-heuristic is formally defined as an iterative generation process, which guides a subordinate heuristic by combining intelligently different concepts for exploring and exploiting the search space, and learning strategies are used to structure information in order to find efficiently near-optimal solutions [5]. It means that two components intensification and diversification should be properly balanced in the search process. The main difference between intensification and diversification is that during the intensification procedure, the search focuses on examining neighbors of elite solutions. The diversification stage, on the other hand, encourages the search process to examine unvisited regions and to generate solutions that differ in various significant ways from those produced before [6, 7].

Actually, the key ideas underlying most NIC approaches are candidate generation, evaluation, selection, and update. The common characteristics of adaptation, learning, and evolution lead to the advantages of derivative-free, intuitive guidelines, flexibility, randomness, robustness, high parallelization property, and global optimization and therefore make the NIC algorithms become the dominant approaches in the field of optimization over the conventional algorithms. However, they also have some distinct differences, and each has its own advantages and drawbacks. The stand-alone NIC methods are still not efficient enough at handling uncertainty and imprecision in practice. Additionally, practicing engineers often face the difficulty of choosing the most suitable NIC methods to meet particular engineering requirements. The capability of overcoming the shortcomings of individual algorithms without losing their advantages makes the hybrid NIC techniques superior to the stand-alone ones.

Firstly proposed by Geem et al. in 2001 [8], the harmony search (HS) method is inspired by the underlying principles of the musicians' improvisation of the harmony. During the recent years, it has been successfully applied in the areas of benchmark function optimization [9], mechanical structure design [10], pipe network optimization [11], magnetic resonance imaging (MRI) segmentation [12], and redundancy optimization problems of electrical and mechanical systems [13]. A lot of modified HS algorithms have been studied in the past decade. For example, based on the continuous HS, Geem [14] proposes a discrete version by introducing the stochastic derivatives for the discrete variables involved. Omran and Mahdavi [15] embed the ideas borrowed from SI into the regular HS and develop an improved HS technique: Global-best HS. Thus, the aim of this book is to give the readers a brief introduction to the HS as well as its hybridization with other NIC algorithms in optimization problem solving.

1.2 Chapter Preview

In this book, this chapter gives a concise introduction to the background of the NIC methods. Chapter 2 presents the overview of the HS including its inspiration and the basic HS optimization algorithm. The HS in context with other NIC algorithms is discussed in Chap. 3. Chapter 4 reviews the existing variations of the HS and its current research trends. Three different hybrid optimization algorithms based on the HS are demonstrated in Chap. 5. Finally, some conclusions and remarks are drawn in Chap. 6.

References

1. V. Ramos, C. Fernandes, A.C. Rosa. A Social Cognitive Maps, Swarm Collective Perception and Distributed Search on Dynamic Landscapes (2005), Available at http://arxiv.org/abs/nlin. AO/0502057
2. G. Beni, U. Wang Swarm intelligence in cellular robotic systems, in *Abstract of the NATO Advanced Workshop on Robots and Biological Systems* (Tuscany, Italy, June 1989)
3. A. Abraham, C. Grosan, V. Ramos (eds.), *Swarm Intelligence and Data Mining* (Springer, Berlin, 2006)
4. S. Kirkpatrick, C. Gelatt, M. Vecchi, Optimization by simulated annealing. Science **220**(4598), 671–680 (1983)
5. I.H. Osman, G. Laporte, Metaheuristics: a bibliography. Ann. Oper. Res. **63**, 513–623 (1996)
6. F. Glover, M. Laguna, *Tabu Search* (Kluwer Academic Publishers, Boston, 1997)
7. C. Blum, A. Roli, Metaheuristics in combinatorial optimization: overview and conceptual comparison. ACM Comput. Surv. **35**(3), 268–308 (2003)
8. Z.W. Geem, J.H. Kim, G.V. Loganathan, A new heuristic optimization algorithm: harmony search. Simulation **76**(2), 60–68 (2001)
9. K.S. Lee, Z.W. Geem, A new meta-heuristic algorithm for continuous engineering optimization: harmony search theory and practice. Comput. Methods Appl. Mech. Eng. **194**(36–38), 3902–3922 (2005)
10. K.S. Lee, Z.W. Geem, A new structural optimization method based on the harmony search algorithm. Comput. Struct. **82**(9–10), 781–798 (2004)
11. Z.W. Geem, J.H. Kim, G.V. Loganathan, Harmony search optimization: application to pipe network design. Int. J. Model. Simul. **22**(2), 125–133 (2002)
12. O.M. Alia, R. Mandava, M.E. Aziz, A hybrid harmony search algorithm for MRI brain segmentation. Evol. Intel. **4**(1), 31–49 (2001)
13. N Nahas, D. Thien-My, Harmony search algorithm: application to the redundancy optimization problem. Eng. Optim. **42**(9), 845–861 (2010)
14. Z.W. Geem, Novel derivative of harmony search algorithm for discrete design variables. Appl. Math. Comput. **199**(1), 223–230 (2008)
15. M.G.H. Omran, M. Mahdavi, Global-best harmony search. Appl. Math. Comput. **198**(2), 643–656 (2008)

Chapter 2
The Overview of Harmony Search

Abstract When musicians compose the harmony, they usually try various possible combinations of the music pitches stored in their memory, which can be considered as a optimization process of adjusting the input (pitches) to obtain the optimal output (perfect harmony). Harmony search draws the inspiration from harmony improvisation, and has gained considerable results in the field of optimization, although it is a relatively NIC algorithm. With mimicking the rules of various combining pitches, harmony search has two distinguishing operators different from other NIC algorithms: harmony memory considering rate (HMCR) and pitch adjusting rate (PAR) that are used to generate and further mutate a solution, respectively. This candidate generation mechanism and single search memory involved decide its excellence in structure simplicity and small initial population. This chapter presents the discussions of the inspiration of harmony search, the basic harmony search optimization algorithm, and an overview of different application areas of the harmony search.

Keywords Harmony search method · Optimization · Hybrid harmony search methods · Benchmarks

2.1 The Inspiration of Harmony Search

The HS was initially proposed by Geem [1] and applied to solve the optimization problem of water distribution networks in 2000. As a novel population-based meta-heuristic algorithm, during the recent years, it has gained great research success in the areas of mechanical engineering, control, signal processing, etc. However, different from most emerging NIC algorithms, the inspiration of the HS is not from the natural phenomena, for example, the CSA is inspired by artificial immune system, and the collective behavior among the unsophisticated individuals of some living creatures has promoted the swarm intelligence, but is conceptualized from the musical process of searching for a perfect state of harmony determined by aesthetic standards.

As we know, when musicians compose the harmony, they usually try various possible combinations of the music pitches stored in their memory. This kind of

© The Author(s) 2015   5
X. Wang et al., *An Introduction to Harmony Search Optimization Method*,
SpringerBriefs in Computational Intelligence,
DOI 10.1007/978-3-319-08356-8_2

Table 2.1 Comparison of harmony improvisation and optimization

Comparison factors	Harmony improvisation	Optimization
Targets	Aesthetic standard	Objective function
Best states	Fantastic harmony	Global optimum
Components	Pitches of instruments	Values of variables
Process units	Each practice	Each iteration

efficient search for a perfect harmony is analogous to the procedure of finding the optimal solutions to engineering problems. The HS method is inspired by the explicit principles of the harmony improvisation [2]. Table 2.1 presents the comparison of harmony improvisation and optimization [3].

2.2 The Basic Harmony Search Algorithm

The music improvisation is a process of searching for the better harmony by trying various combinations of pitches that should follow any of the following three rules [2]:

1. playing any one pitch from the memory;
2. playing an adjacent pitch of one pitch from the memory;
3. playing a random pitch from the possible range.

This process is mimicked in each variable selection of the HS algorithm. Similarly, it should follow any of the three rules below:

1. choosing any value from the HS memory;
2. choosing an adjacent value from the HS memory;
3. choosing a random value from the possible value range.

The three rules in the HS algorithm are effectively directed using two essential parameters: harmony memory considering rate (HMCR) and pitch adjusting rate (PAR). Figure 2.1 shows the flowchart of the basic HS method, in which there are four principal steps involved.

Step 1. Initialize the HS memory (HM). The initial HM consists of a given number of randomly generated solutions to the optimization problems under consideration. For an n-dimension problem, an HM with the size of HMS can be represented as follows:

$$HM = \begin{bmatrix} x_1^1, x_2^1, \ldots, x_n^1 \\ x_1^2, x_2^2, \ldots, x_n^2 \\ \vdots \\ x_1^{HMS}, x_2^{HMS}, \ldots, x_n^{HMS} \end{bmatrix}, \qquad (2.1)$$

where $\left[x_1^i, x_2^i, \ldots, x_n^i \right]$ $(i = 1, 2, \ldots, HMS)$ is a solution candidate. HMS is typically set to be between 50 and 100.

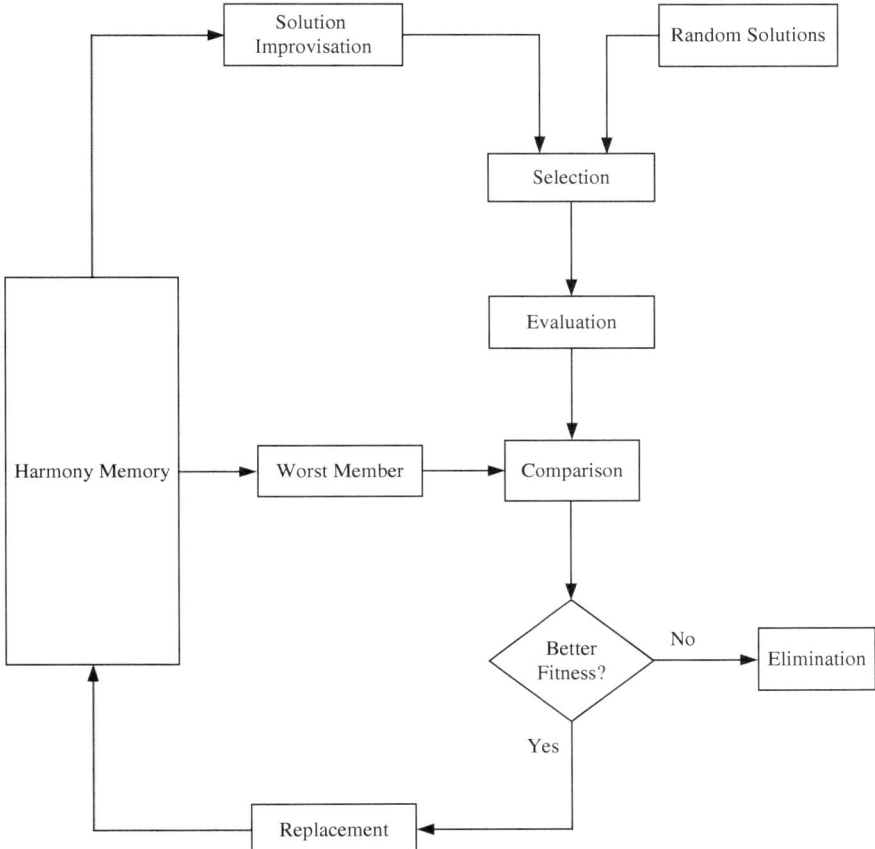

Fig. 2.1 Harmony Search (HS) method

Step 2. Improvise a new solution $\left[x'_1, x'_2, \ldots, x'_n\right]$ from the HM. Each component of this solution, x'_j, is obtained based on the HMCR. The HMCR is
defined as the probability of selecting a component from the present
HM members, and 1-HMCR is, therefore, the probability of generating
it randomly. If x'_j comes from the HM, it is chosen from the jth
dimension of a random HM member, and it can be further mutated
according to the PAR. The PAR determines the probability of a candidate from the HM to be mutated. Obviously, the improvisation of
$\left[x'_1, x'_2, \ldots, x'_n\right]$ is rather similar to the production of the offspring in the
genetic algorithm (GA) [4, 5] with the mutation and crossover oper-
ations. However, the GA creates fresh chromosomes using only one
(mutation) or two (simple crossover) existing ones, while the

generation of new solutions in the HS method makes full use of all the HM members.

Step 3. Update the HM. The new solution from Step 2 is evaluated. If it yields a better fitness than that of the worst member in the HM, it will replace that one. Otherwise, it is eliminated.

Step 4. Repeat Step 2 to Step 3 until a preset termination criterion, e.g., the maximal number of iterations, is met.

Apparently, the HMCR and PAR are two basic parameters in the HS algorithm, which control the component of solutions and even affect convergence speed. The former is used to set the probability of utilizing the historic information stored in the HM. For example, 0.9 indicates that each component of a new solution will be chosen from the HM with 90 % probability, and 10 % probability from the entire feasible range. Each component of the solution is subject to whether it should be pitch-adjusted, which is determined by PAR. 1-PAR means the rate of doing nothing. For example, a PAR of 0.3 indicates that the neighboring value will be chosen with 30 % probability.

Similar to the GA, particle swarm optimization (PSO) [6–8], and differential evolution (DE) [9, 10], the HS method is a random search technique. It does not require any prior domain information, such as the gradient of the objective functions. However, different from those population-based evolutionary approaches, it only utilizes a single search memory to evolve. Therefore, the HS method has the characteristics of algorithm simplicity. On the other hand, it occupies some inherent drawbacks, e.g., weak local search ability. In the following two chapters, the comparisons between the HS and other NIC optimization algorithms and the variations of the HS will be discussed individually.

2.3 Survey of the Harmony Search Applications

In the real world, modern science and industry are indeed rich in the problems of optimization. Since the HS was originally proposed by Geem [11] and applied to solve the optimization problem of water distribution networks in 2000, the applications of the HS have covered many areas including industry, optimization benchmarks, power systems, medical science, control systems, construction design, and information technology [12].

2.3.1 Optimization Benchmarks

Optimization benchmarks for the hybridization of the HS method with other approaches are one principal application area. Different variants based on the HS have been demonstrated their improvement and efficiency through various benchmark functions. Combined with semantic genetic operators, Castelli et al. [13] propose a geometric selective harmony search (GSHS) method with three main differences from the original HS: (1) the memory consideration process involves the presence of a selection procedure, (2) the algorithm integrates a particular recombination operator that combines the information of two harmonies, and (3) the algorithm utilizes a mutation operation that uses the PAR parameter. Therefore, geometric semantic crossover produces offspring that is not worse than the worst of its parents, and geometric semantic mutation causes a perturbation on the semantics of solutions, whose magnitude is controlled by a parameter. Five different HS algorithms have been compared using 20 benchmark problems, and the GSHS outperforms the others with statistically significant enhancement in almost all the cases.

2.3.2 Industry

Industry is a prominent area full of various practical optimization issues subject to multi-modal, constrained, nonlinear, and dynamical. The HS algorithm proposed by Saka [14] determines the optimal steel section designations from the available British steel section table, and implements the design constraints from BS5950. Recently, an enhanced harmony search (EHS) in [15] is developed enabling the HS algorithm to quickly escape from local optima. The proposed EHS algorithm is utilized to solve four classical weight minimization problems of steel frames including two-bay, three-storey planar frame subject to a single-load case, one-bay, ten-storey planar frame consisting of 30 members, three-bay, twenty four-storey planar frame, and Spatial 744 member steel frame. In [16], the HS is used to select the optimal parameters in the tuned mass dampers [16]. Fesanghary et al. [17] propose a hybrid optimization method based on the global sensitivity analysis and HS for the optimal design of shell and tube heat exchangers.

2.3.3 Power Systems

There is a lot of work focused on the optimization issues concerning power systems, such as cost minimization. A modified HS algorithm is proposed to handle non-convex economic load dispatch of real-world power systems. The economic load dispatch and combined economic and emission load dispatch problems can be converted into the minimization of the cost function [18]. Sinsuphan et al. [19] combine the HS with sequential quadratic programming and GA to solve the optimal power flow problems. The objective function to be optimized is the total generator fuel costs in the entire system. The chaotic self-adaptive differential HS algorithm, proposed by Arul et al. [20], is employed to deal with the dynamic economic dispatch problem.

2.3.4 Signal and Image Processing

Li and Duan [21] modify the HS by adding a Gaussian factor to adjust the bandwidth (bw). With this modified HS, they develop a pre-training process to select the weights used in the combining of feature maps to make the target more conspcuity in the saliency map. In their method based on the HS, Fourie et al. [22] design a harmony filter using the improved HS algorithm for a robust visual tracking system.

2.3.5 Others

In addition to the aforementioned applications, the HS has also been widely employed in a large variety of fields, including transportation, manufacturing, robotics, control, and medical science [11]. Many traffic modeling software are capable of finding the optimal or near-optimal signal timings using different optimization algorithms. For example, Ceylan [23] proposes a modified HS with embedded hill climbing algorithm for further tuning the solutions in the stochastic equilibrium network design. The modified HS algorithm is also used in parameter identification of the solar cell mathematical models [24]. Miguel et al. [25] employ the HS in damage detection under the ambient vibration.

References

1. Z.W. Geem, Optimal cost design of water distribution networks using harmony search, Dissertation, Korea University (2000)
2. Z.W. Geem, J.H. Kim, G.V. Loganathan, A new heuristic optimization algorithm: harmony search. Simulation **76**(2), 60–68 (2001)
3. X. Wang, Hybrid Nature-Inspired Computation Methods for Optimization, Dissertation, Helsinki University of Technology (2009)
4. R. Poli, W.B. Langdon, *Foundations of Genetic Programming* (Springer, Berlin, 2002)
5. M. Krug, S.K. Nguang, J. Wu et al., GA-based model predictive control of boiler-turbine systems. Int. J. Innov. Comput. Inf. Control **6**(11), 5237–5248 (2010)
6. A.P. Engelbrecht, *Fundamentals of Computational Swarm Intelligence* (Wiley, West Sussex, 2005)
7. C.T. Lin, J.G. Wang, S.M. Chen, 2D/3D face recognition using neural network based on hybrid Taguchi-particle swarm optimization. Int. J. Innov. Comput. Inf. Control **7**(2), 537–553 (2011)
8. X. Cai, Z. Cui, J. Zeng et al., Particle swarm optimization with self-adjusting cognitive selection strategy. Int. J. Innov. Comput. Inf. Control **4**(4), 943–952 (2008)
9. R. Storn, K. Price, Differential evolution: a simple and efficient adaptive scheme for global optimization over continuous spaces. J. Glob. Optim. **11**, 341–359 (1997)
10. V. Vegh, G.K. Pierens, Q.M. Tieng, A variant of differential evolution for discrete optimization problems requiring mutually distinct variables. Int. J. Innov. Comput. Inf. Control **7**(2), 897–914 (2011)

11. Z.W. Geem (ed.), *Music-Inspired Harmony Search Algorithm* (Springer, Berlin, 2001)
12. D. Manjarresa, I. Landa-Torresa, S. Gil-Lopeza et al., A survey on applications of the harmony search algorithm. Eng. Appl. Artif. Intel. **26**(8), 1818–1831 (2013)
13. M. Castelli, S. Silva, L. Manzoni et al., Geometric selective harmony search. Inf. Sci. (2014). doi:10.1016/j.ins.2014.04.001
14. M.K. Saka, Optimum design of steel skeleton structures, in *Music-Inspired Harmony Search Algorithm*, ed. by Z.W. Geem (Springer, Berlin, 2009), pp. 87–112
15. R. Mahmoud, M. Maheri, M. Narimani, An enhanced harmony search algorithm for optimum design of side sway steel frames. Comput. Struct. **136**, 78–89 (2014)
16. G. Bekdaş, S.M. Nigdeli, Estimating optimum parameters of tuned mass dampers using harmony search. Eng. Struct. **33**(9), 2716–2723 (2011)
17. M. Fesanghary, E. Damangir, I. Soleimani, Design optimization of shell and tube heat exchangers using global sensitivity analysis and harmony search algorithm. Appl. Therm. Eng. **29**(5–6), 1026–1031 (2009)
18. B. Jeddi, V. Vahidinasab, A modified harmony search method for environmental/economic load dispatch of real-world power systems. Energy Convers. Manag. **78**, 661–675 (2014)
19. N. Sinsuphan, U. Leeton, T. Kulworawanichpong, Optimal power flow solution using improved harmony search method. Appl. Soft Comput. **13**(5), 2364–2374 (2013)
20. R. Arul, G. Ravi, S. Velusami, Chaotic self-adaptive differential harmony search algorithm based dynamic economic dispatch. Int. J. Electr. Power Energy Syst. **50**, 85–96 (2013)
21. J. Li, H. Duan, Novel biological visual attention mechanism via Gaussian harmony search. Optik-Int. J. Light Electron Opt. **125**(10), 2313–2319 (2014)
22. J. Fourie, S. Mills, R. Green, Harmony filter: a robust visual tracking system using the improved harmony search algorithm. Image Vis. Comput. **28**(12), 1702–1716 (2010)
23. H. Ceylan, H. Ceylan, A hybrid harmony search and TRANSYT hill climbing algorithm for signalized stochastic equilibrium transportation networks. Transp. Res. Part C Emerg. Technol. **25**, 152–167 (2012)
24. A. Askarzadeh, Parameter identification for solar cell models using harmony search-based algorithms. Sol. Energy **86**(11), 3241–3249 (2012)
25. L.F.F. Miguel, L.F.F. Miguel, J.J. Kaminski et al., Damage detection under ambient vibration by harmony search algorithm. Expert Syst. Appl. **3**(10), 9704–9714 (2012)

Chapter 3
The Harmony Search in Context with Other Nature Inspired Computational Algorithms

Abstract Inspiration drawn from nature and modeling of natural processes are the two common characteristics existing in most NIC algorithms. These methodologies, therefore, share many similarities, e.g., adaptation, learning, and evolution, and have a general flowchart including candidate initialization, operation, and renewal. On the other hand, mimicking various natural phenomena leads to their different generation, evaluation, selection, and update mechanisms, which may result in individual inherent distinctive properties, advantages, as well as drawbacks in the performances of dealing with different optimization problems. For example, the CSA on the basis of modeling the clonal selection principle of the artificial immune system performs well in the local search but suffers from a long convergence time. This chapter compares three typical evolutionary optimization methods, GA, CSA, and HS, with regard to their structures and performances using illustrative examples.

Keywords Harmony search method · Genetic algorithm · Clonal selection algorithm · Optimization

3.1 Genetic Algorithm

According to Darwin's natural selection and survival of the fittest theory in the book of 'Origin of Species,' competition exists among the individuals for the limited resources, and affects their mating and reproduction. This elitist strategy promotes the most adaptive genes to be kept in the successive generation and further to create better offspring that are suitable to the environment. As the most widely used branch of evolutionary search methods, the genetic algorithm (GA) was firstly developed by John Holland in the 1960s [1]. To mimic the genetic process of biological organisms, it needs to represent a solution to the given problem as a chromosome. The new solutions in the sequent generations are subject to two types of genetic operators: mutation and crossover. The flowchart of

© The Author(s) 2015
X. Wang et al., *An Introduction to Harmony Search Optimization Method*,
SpringerBriefs in Computational Intelligence,
DOI 10.1007/978-3-319-08356-8_3

the basic GA is illustrated in Fig. 3.1, and the iteration steps are interpreted as the following [2–4].

Step 1   Choose an appropriate encoding method and evaluation function for possible solutions. The encoding method transforms the parameter space of solutions into bit-string representations. A bit-string is named 'chromosome' in the GA. The evaluation function is always case dependent and can be applied to evaluate different solution candidates.

Step 2   Initialize a population of candidates (chromosomes) randomly.

Step 3   Evaluate these chromosomes and calculate their fitnesses based on the evaluation function in Step 1.

Step 4   Create new chromosomes by applying the mutation and crossover operators on the current population. Mutation is an operator that introduces variations into the chromosomes and thus adds the property of random search to the GA. The mutation operator acts in such a way that each bit of the bit-strings is replaced with a randomly generated bit on a probability bias, P_m. Crossover is a unique GA operator that combines two parent chromosomes to produce offspring containing subparts from both parents. A probability term, P_c, determines its operation rate. As a matter of fact, the aforementioned two probabilities for mutation and crossover operators, P_m and P_c, play an important role in the convergence process of the GA.

Step 5   Evaluate the new chromosomes produced in Step 4, and calculate the corresponding fitnesses as in Step 3.

Step 6   Combine the chromosomes from both Step 2 and Step 4 together, and select the ones with top fitnesses in order to keep the population scale always fixed in each generation. In other words, only the mostly fitted chromosomes have the possibility of survival.

Step 7   Repeat Step 4 to Step 7 until a preset criterion of the evaluation function is met.

Apparently, the reproduction of the GA depends on two operators: crossover and mutation that can be considered as two approaches to exploring a cost surface. The former is a process of generating offspring through the recombination of genes from the selected parents. The latter can introduce traits not in the initial population and prevent the process from converging too fast. During the overall process of the GA, the best chromosomes enter the next generation without being mutated, which means the more chromosomes survive in the following iteration, the less the diversity of the new population. This selection mechanism, called elitism, is one distinguishing property of the GA.

Fig. 3.1 Flowchart of basic
genetic algorithm (GA)

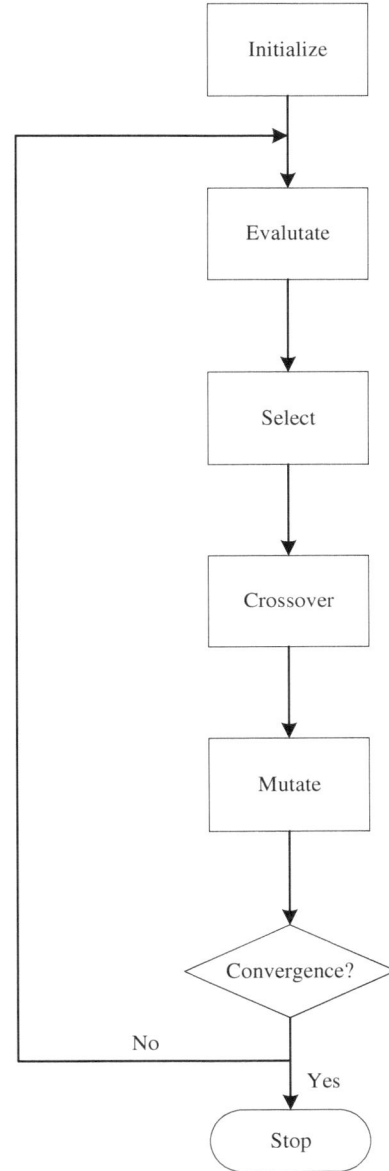

3.2 Clonal Selection Algorithm

Inspired by the clonal selection principle (CSP), the clonal selection algorithm
(CSA) has been studied and applied to cope with demanding optimization problems, due to its superior search capability compared with the classical optimization

Fig. 3.2 Flowchart of basic
clonal selection algorithm
(CSA)

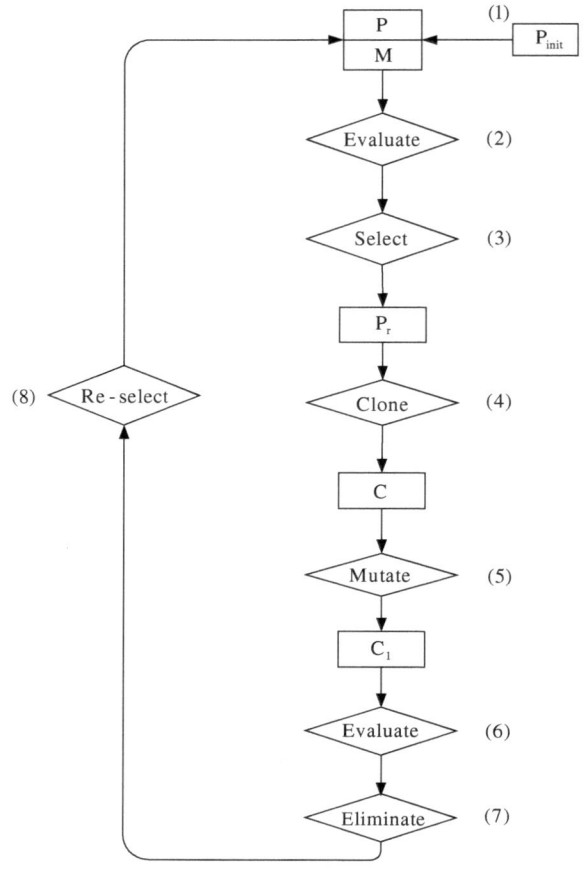

techniques [5]. The CSP explains how an immune response is mounted, when a non-self antigenic pattern is recognized by the B cells. In the natural immune systems, only the antibodies that can recognize the intruding antigens are selected to proliferate by cloning [6]. Hence, the fundamental idea of the CSA is that those cells (antibodies) capable of recognizing the non-self cells (antigens) will proliferate. To summarize, the main ideas of the CSA borrowed from the CSP are as follows [7]:

1. maintenance of memory cells functionally disconnected from the repertoire;
2. selection and cloning of the most stimulated antibodies;
3. affinity maturation and reselection of clones with higher affinity; and
4. mutation rate proportional to cell affinity.

The flowchart of a basic CSA is shown in Fig. 3.2, and it involves the following nine iteration steps [8].

1. Initialize the antibody pool P_{init} including the subset of memory cells (M).
2. Evaluate the fitnesses of all the antibodies (affinity with the antigen) in population P.
3. Select the best candidates (P_r) from population P, according to their fitnesses.
4. Clone P_r into a temporary antibody pool (C).
5. Generate a mutated antibody pool (C_1). The mutation rate of each antibody is inversely proportional to its fitness.
6. Evaluate all the antibodies in C_1.
7. Eliminate those antibodies similar to the ones in C, and update C_1.
8. Reselect the antibodies with better fitnesses from C_1 to construct memory set M. Other improved individuals of C_1 can replace certain existing members with poor fitnesses in P to maintain the whole antibody diversity.
9. Return back to Step 2, if the preset performance criteria are not met. Otherwise, terminate.

We emphasize that a unique mutation operator is used in Step 5, in which the mutated values of the antibodies are inversely proportional to their fitnesses by means of choosing different mutation variations. That is to say, the better fitness the antibody has, the less it may change. The similarity among the antibodies can also affect the overall convergence speed of the CSA. Thus, the strategy of antibody suppression inspired by the immune network theory [9] is introduced to eliminate the newly generated antibodies, which are too similar to those already in the candidate pool (Step 7). With such a diverse antibody pool, the CSA can effectively avoid being trapped into the local minima and provide the optimal solutions to the multimodal problems [10]. In summary, the antibody cloning and fitness-related mutation are the two remarkable characteristics of the CSA.

3.3 Performance Comparisons Among HS, GA, and CSA

Similar to most population-based algorithms, the GA, CSA, and HS mimic the natural phenomena and physical principles, and they, therefore, share the same general structure including initialization, operation, and renewal, which brings out their similarities in adaptation, learning, and evolution during the optimization process. The general evolving procedure can be explained as follows. First, randomly generate an initial collection of candidate solutions. After that, produce new members by making changes to the selected candidates and then evaluate them. The changes may involve merging two or more existing members together or introducing random variations to the current candidates. Finally, replace those solutions that are outdated with the improved ones. Hence, the key ideas underlying these NIC algorithms are candidate generation, evaluation, selection, and population update.

Due to their natural inspirations, different candidate renewal mechanisms decide their inherent characteristics including both advantages and drawbacks.

Search range	Global optima	n
$[-10, 10]$	$f(x) = 0$	20

For example, in the HS, based on the PAR, the new candidates improvised from the existing HM members may go through a mutation procedure. The mutation is generally a 'blind' and local exploration in the search space. Additionally, the candidates selection and update highly depend on the past search experience, which leads to the outdated information stored in the harmony memory.

As a simple and illustrative example, minimization of Function 3.1, is employed here to demonstrate the comparison performances of GA, CSA, and HS. The details of this function are given in Table 3.1.

Function 3.1:

$$f(x) = \sum_{i=1}^{n} 100 \left(x_{i+1} - x_i^2 \right)^2 + (x_i - 1)^2. \tag{3.1}$$

Figure 3.3 illustrates the average convergence procedures of the GA, CSA, and HS over 50 runs in the optimization of Function 3.1, which are represented by the dash-dotted, dash, and solid lines, respectively. The initial population contains 20

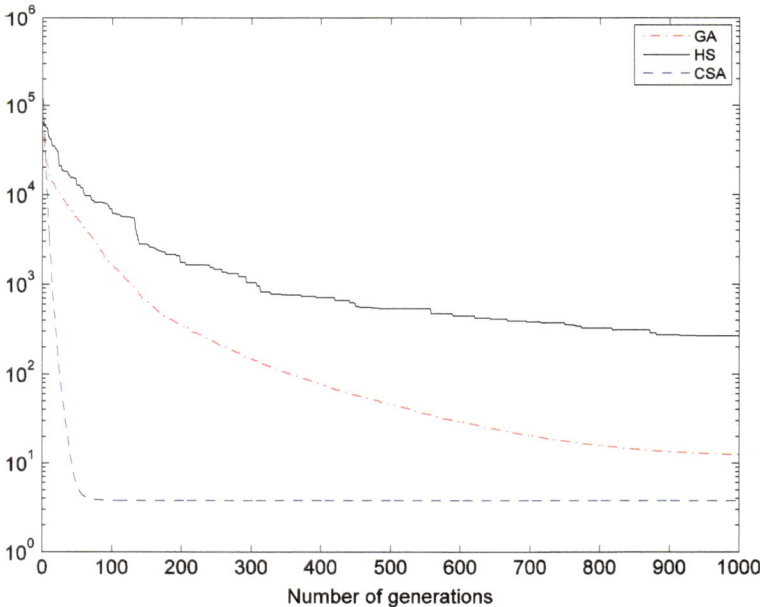

Fig. 3.3 Convergence procedures of CSA, GA, and HS in optimization of Function 3.1 (20 candidates)

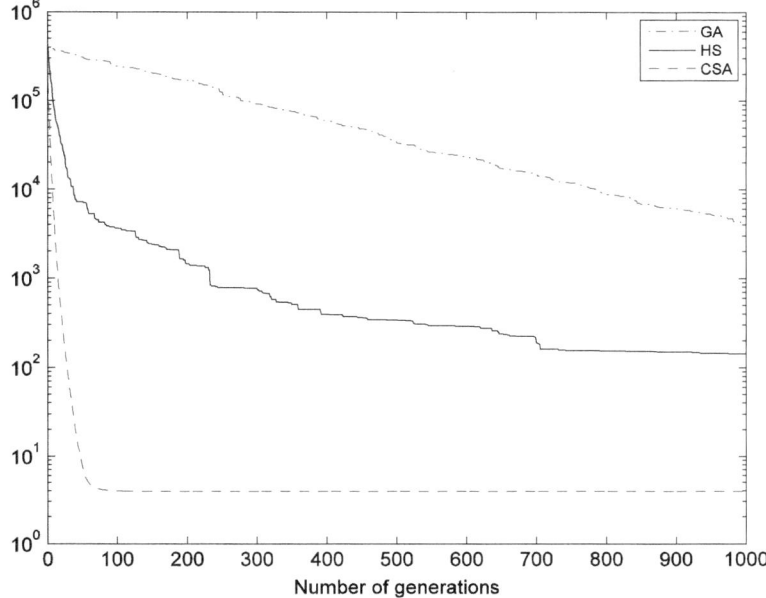

Fig. 3.4 Convergence procedures of CSA, GA, and HS in optimization of Function 1 (10 candidates)

candidates in this simulation. Apparently, the convergence speed of the HS is the lowest, and the CSA is the fastest among these three algorithms.

We also run these three methods with a half initial population size (10 candidates) for another comparison shown in Fig. 3.4. In Fig. 3.4, the convergence speed of the GA is the lowest, and the CSA is still the fastest among the three algorithms. The performance of the HS is comparatively better this time, which means that it can handle the optimization problems with even a small number of candidates used.

We compare the simulation time (in seconds) under different conditions including function dimensions and numbers of candidates given in Table 3.2. Their average simulation time over 30 runs is shown after 1,000 generations. Obviously, the HS uses the shortest simulation time, although it has a slow convergence speed with regard to the number of generations shown in Figs. 3.3 and 3.4. The main

Table 3.2 The comparison of average simulation time among CSA, GA, and HS

Dimension	Number of candidates	GA (s)	HS (s)	CSA (s)
10	10	0.1238	0.0796	0.5128
10	20	0.1456	0.0850	0.5324
30	10	0.1259	0.0810	0.5898
30	20	0.1498	0.0861	0.6410

reason why the CSA occupies the longest time among these three algorithms is due to the affinity measure of antibodies for eliminating those similar candidates and maintaining the diversity of the candidate pool. The simple computation and structure of the HS makes it widely employed in the hybridization with other NIC algorithms for different optimization problem solving.

References

1. J.H. Holland, *Adaptation in natural and artificial systems* (University of Michigan Press, Ann Arbor, 1975)
2. K.F. Man, K.S. Tang, S. Kwong, Genetic algorithms: Concepts and applications. IEEE Trans. Ind. Electron. **43**(5), 519–534 (1996)
3. K.S. Tang, K.F. Man, S. Kwong et al., Genetic algorithms and their applications. IEEE Signal Process **6**, 22–37 (1996)
4. X.Z. Gao, S.J. Ovaska, Genetic algorithm training of Elman neural network in motor fault detection. Neural Comput. Appl. **11**(1), 37–44 (2002)
5. X Wang, X.Z. Gao, S.J Ovaska, Artificial immune optimization methods and applications-a survey. in *IEEE International Conference on Systems, Man, and Cybernetics*, The Hague, The Netherlands, 10–13 Oct 2004
6. J. Timmis, P. Andrews, N. Owens et al., An interdisciplinary perspective on artificial immune systems. Evolut. Intell. **1**(1), 2–26 (2008)
7. L.N. Castro, F.J. Zuben, Learning and optimization using the clonal selection principle. IEEE Trans. Evolut. Comput. **6**(3), 239–251 (2002)
8. X Wang, Clonal selection algorithm in power filter optimization. in *IEEE Mid-Summer Workshop on Soft Computing in Industrial Applications*, Espoo, Finland, 28–30 June 2005
9. D Dasgupta, Advances in artificial immune systems. IEEE Comput. Intell. Mag. **1**(4), 40–49 (2006)
10. X Wang, X.Z. Gao, S.J. Ovaska, A novel particle swarm-based method for nonlinear function optimization. Int. J. Comput. Intell. Res. **4**(3), 281–289 (2008)

Chapter 4
The Variations of Harmony Search and Its Current Research Trends

Abstract Since the HS was firstly assessed and utilized in the optimal cost design of water distribution network in 2006, it has attracted growing research interest in handling a large variety of optimization problems. As aforementioned, the original HS has its own inherent drawbacks (slow convergence and outdated information). Hence, a lot of relevant literature proposes many variations of the HS, which can be classified into two major categories: modification based on the regular HS and hybridization with other NIC algorithms. The former targets the improvement of parameters or operators of the HS to enhance the quality of solutions, while the latter incorporates some optimization approaches to overcome its drawbacks and improve the performance. From the viewpoint of structure, the cooperation in the fusion strategies of the hybrid HS methods could be in the manner of either cooperator or embedded operators. This chapter first gives an overview of the variations of the HS and the current research trends, and a modified HS method for constrained optimization is next discussed in details.

Keywords Harmony search method · Constrained optimization · Adaptive parameters · Hybrid harmony search methods · Parameter tuning · Penalty function

4.1 Modification Based on the Original Harmony Search

4.1.1 Revised Population Update Mechanism

To effectively maintain the diversity of the HM members, the qualification of a solution candidate as a new HM member should be based on not only its fitness but also its similarity to the existing members. A HS member control mechanism is introduced to decide whether the new solution will replace the worst member or not [1]. In more details, the newly generated member will replace the worst one of the HM, if it meets the three conditions:

1. the number of the HM member in the vicinity is smaller than a preset threshold,
2. its fitness is better than the average fitness of the HM members in the vicinity,

© The Author(s) 2015   21
X. Wang et al., *An Introduction to Harmony Search Optimization Method*,
SpringerBriefs in Computational Intelligence,
DOI 10.1007/978-3-319-08356-8_4

3. it is better than the worst HM member.

To deal with the dynamic optimization problems, the modified HS method proposed by Turky divides a population into several subpopulations that occupy different areas in the search space to either intensify or diversify the search process. Each subpopulation represents one HS, and they interact with one another by merging all the subpopulations together and partitioning them again when a change in the environment is detected [2].

4.1.2 Modified Pitch Adjustment Strategy

The original HS algorithm uses fixed values for both PAR and bw, which are set in the initialization, and cannot be changed during new generations. In [4], a new HS algorithm uses variable PAR and bw in the improvisation step as follows:

$$PAR(gn) = PAR_{min} + \frac{PAR_{max} - PAR_{min}}{NI} \times gn, \qquad (4.1)$$

$$bw(gn) = bw_{max}exp\left[\frac{Ln\left(\frac{bw_{min}}{bw_{max}}\right)}{NI} \cdot gn\right], \qquad (4.2)$$

where NI is the number of solution vector generations, and gn is the generation number. The small bw values in the final generations increase the fine-tuning of solution vectors, but in the early generations, bw must take a bigger value to enforce the HS algorithm to increase the diversity of solution vectors. Furthermore large PAR values with small bw values usually lead to the improvement of the best solutions in final generations so that the HS algorithm converges to the optimal solution vector. The key difference from the original HS is that PAR and bw change dynamically with generation numbers [4].

In Degertekins' self-adaptive HS algorithm, bw is removed and replaced by the following:

$$x_i^{new} = x_i^{new} + \left[max(HM)_i - x_i^{new}\right] \times u(0, 1) \quad if \ u(0, 1) \leq 0.5, \qquad (4.3)$$

$$x_i^{new} = x_i^{new} - \left[x_i^{new} - min(HM)_i\right] \times u(0, 1) \quad if \ u(0, 1) > 0.5, \qquad (4.4)$$

where $min(HM_i)$ and $max(HM_i)$ are the lowest and highest values of the ith design variable in the HM. $u(0, 1)$ is a uniform random number in the [0, 1] range [3].

In the modified HS proposed by Fesangharya et al. [5], PAR and bw are also changed with the iterations. This improved HS is used in the optimal design of shell and tube heat exchangers, and can yield a better performance than the GA.

In [6], an improved global-best harmony search is proposed. In the pitch generation scheme, a randomly selected pitch is adjusted according to a Gaussian distribution in the pitch selection step, and the current pitch of the best harmony member is selected and adjusted at the same time using the uniform distribution instead in the pitch adjustment step.

Kumar et al. [7] propose a parameter adaptive HS algorithm, called PAHS, in which PAR is changed exponentially and HMCR is changed linearly during the process of improvisation. In addition, they compare different combinations of linearity or exponent based HMCR and PAR. The simulation results demonstrate that the PAHS is able to provide the best results in high-dimensional and noisy benchmark and data clustering problems.

4.2   Hybridization with Other NIC Algorithms

Actually, the most popular variations of the HS draw their inspiration from the fusion with other NIC algorithms rather than only modifying the operators. In such a heuristic structure, the hybridization might fuse two or even more types of NIC algorithms in the form of either cooperator or embedded operator. In the former manner, the common information and knowledge is exchanged and shared among the algorithms during the search process, and the latter can be characterized by the hybridization architecture of merging one NIC algorithm into the other.

4.2.1   Cooperator

Proposed by Omran, the global-best HS introduces the concept of swarm intelligence and modifies the pitch adjustment step of the HS such that the new solution is affected by the best harmony in the harmony memory. The essence of this new method is replacing the bw parameter and adding a social dimension to the HS [8]. Amini and Ghaderi design an AntHS algorithm based on the fusion of the ACO and HS for the optimal locating of structural dampers. The pheromone and heuristic values borrowed from the ACO are used to influence the search procedure of the HS [9].

4.2.2   Embedded Operator

In the parallel chaotic optimization algorithm (PCOA) proposed by Pan et al. [10], the first stage is to apply the PCOA with twice carrier wave parallel chaos search for global search, and the second stage is to use the HS algorithm for accurate local search. In the heuristic particle swarm optimizer, only the harmony memory

Table 4.1 Variations of HS method with applications

Algorithms	References	Applications
GA − HS	[13, 25]	Power systems
SA + HS	[14]	Power systems
ABC + HS	[15]	Benchmark problems
ACO + HS	[9]	Design
NN − HS	[16]	Robotics
	[26]	Classification
	[27]	Economics
DE + HS	[1]	Benchmark problems
	[28]	Robotics
PSO + HS	[11]	Design
	[24]	Benchmark problems

concept, excluding the HMCR and PAR, is used in the PSO algorithm so as to avoid the search being trapped into local optima in dealing with the variable constraints [11].

In fact, more and more HS variations are based on both the hybridization and modification of the HS. For example, the HMCR and PAR are excluded from a novel global HS algorithm, proposed by Zou et al., and replaced by the genetic mutation probability. In addition, the improvisation step of the HS is modified using the idea of social interaction from the PSO [12]. Table 4.1 gives the variations of the HS method as well as their applications.

4.3 A Modified HS Method for Constrained Optimization

4.3.1 Constrained Optimization Problems

Most of the practical optimization problems are indeed constrained optimization problems, whose goal is to find the optimal solution that satisfies a set of given constraints. In general, a constrained optimization problem is described as follows [17, 18]:

Find $\vec{x} = (x_1, x_2, \ldots, x_n)$ to minimize $f(\vec{x})$,

subject to $g_i(\vec{x}) \leq 0$, $i = 1, 2, \ldots, M$ and $h_j(\vec{x}) = 0$, $j = 1, 2, \ldots, N$, where $f(\vec{x})$ is the objective function, and $g_i(\vec{x})$ and $h_j(\vec{x})$ are the inequality and equality constraint functions, respectively. As a matter of fact, the equality constraint functions can be easily transformed into the inequality constraint functions:

$$\left| h_j(\vec{x}) \right| - \varepsilon \leq 0, \tag{4.3}$$

where ε is a small enough tolerance parameter. Therefore, we only consider the inequality constraint functions $g_i(\vec{x}) \leq 0$, $i = 1, 2, \ldots, M$ in this section.

The constrained optimization problems are generally difficult to deal with, because the constraint functions can divide the whole search space into some disjoint islands. Numerous constraint-handling techniques have been proposed and investigated during the past decades [19–22]. One popular solution is to define a new fitness function $F(\vec{x})$ to be optimized [23]. $F(\vec{x})$ is the combination of the objective function $f(\vec{x})$ and weighted penalty terms $P_i(\vec{x})$, $i = 1, 2, \ldots, M$, which reflect the violation of the constraint functions:

$$F(\vec{x}) = f(\vec{x}) + \sum_{i=1}^{M} w_i P_i(\vec{x}), \qquad (4.4)$$

where w_i ($i = 1, 2, \ldots, M$) are the preset weights. The overall optimization performance depends on the penalty terms and their weights, and may significantly deteriorate with inappropriately chosen ones. Unfortunately, there is no analytic way yet to find the best $P_i(\vec{x})$ and w_i. In this section, on the basis of the Pareto dominance, we propose a modified HS method for the direct handling of the given constraints.

4.3.2 A Modified HS Method for Constrained Optimization

It is well known that the regular HS method is not effective in attacking the constrained optimization problems. As aforementioned, the HM only stores the feasible solution candidates. The new HM members are generated either from the existing HM members or in a random way. Nevertheless, they are not guaranteed to always meet all the constraints. Figure 4.1 shows that in the HS method, the new HM members satisfying the constraints can be obtained based on only *trial and error*, which may lead to a time-consuming procedure, especially in case of multiple and complex constraint functions [4].

In our modified HS method, we take advantage of those HM members that do not even meet the constraints. The key issue is how to rank the HM members according to their objective as well as constraint functions values. Here, the values of the constraint functions of all the HM members are stored together with their objective functions values in the HM:

$$\mathrm{HM} = \begin{bmatrix} x_1^1, x_2^1, \ldots, x_n^1, g_1(\vec{x}^1), g_2(\vec{x}^1), \ldots, g_M(\vec{x}^1) \\ x_1^2, x_2^2, \ldots, x_n^2, g_1(\vec{x}^2), g_2(\vec{x}^2), \ldots, g_M(\vec{x}^2) \\ \vdots \\ x_1^{\mathrm{HMS}}, x_2^{\mathrm{HMS}}, \ldots, x_n^{\mathrm{HMS}}, g_1(\vec{x}^{\mathrm{HMS}}), g_2(\vec{x}^{\mathrm{HMS}}), \ldots, g_M(\vec{x}^{\mathrm{HMS}}) \end{bmatrix}, \qquad (4.5)$$

Fig. 4.1 Generation of new HM members satisfying constraints in original HS method

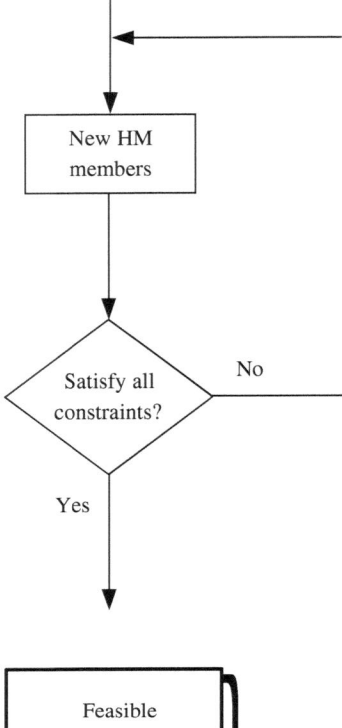

Fig. 4.2 Harmony memory with feasible and infeasible members

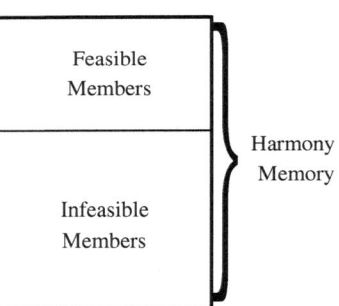

where $\vec{x}^1 = [x_1^1, x_2^1, \ldots, x_n^1]$, $\vec{x}^2 = [x_1^2, x_2^2, \ldots, x_n^2]$, ..., $\vec{x}^{HMS} = [x_1^{HMS}, x_2^{HMS}, \ldots, x_n^{HMS}]$. The HM members are divided into two different parts: feasible members and infeasible members, as illustrated in Fig. 4.2. The former satisfy all the constraint functions, while the latter do not. Thus, the ranking of the HM members is separated into two consecutive stages: ranking of the feasible HM members and ranking of the infeasible ones. The ranking of the feasible HM members is straightforward: They can be sorted using their objective functions values. However, for the infeasible ones, the ranking is based on the Pareto dominance of these HM members [14]. An infeasible HM member dominates another, if none of its constraint functions values is larger and at least one is smaller. Formally, the Pareto dominance is defined as follows. Suppose there are two infeasible HM members, $\vec{x}^1$ and $\vec{x}^2$. If

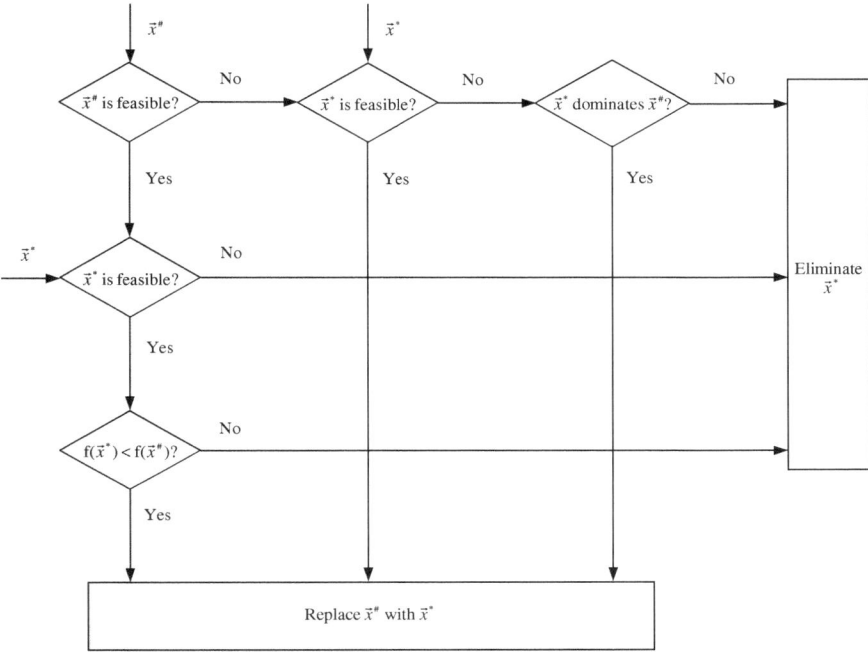

Fig. 4.3 Comparison between $\vec{x}^{\#}$ and $\vec{x}^{*}$, replacement of $\vec{x}^{\#}$ with $\vec{x}^{*}$, and elimination of $\vec{x}^{*}$

$$\forall i \in \{1, 2, \ldots, M\} : g_i(\vec{x}^1) \leq g_i(\vec{x}^2) \wedge \exists i \in \{1, 2, \ldots, M\} : g_i(\vec{x}^1) < g_i(\vec{x}^2),$$

we conclude that $\vec{x}^1$ dominates $\vec{x}^2$. For each infeasible HM member, we can calculate the number of the others that dominate it, which implies its relative degree
of violation of the constraint functions. That is, the rank of an infeasible HM
member is determined by the number of other infeasible HM members by which it
is dominated.

After the whole HM has been ranked, the worst HM member $\vec{x}^{\#}$ can be selected
and compared with the new solution candidate $\vec{x}^{*}$. Note, $\vec{x}^{*}$ does not need to be
feasible. When $\vec{x}^{\#}$ is compared with $\vec{x}^{*}$, $\vec{x}^{*}$ will replace $\vec{x}^{\#}$ only in one of the
following three cases:

1. $\vec{x}^{*}$ is feasible, and $\vec{x}^{\#}$ is infeasible,
2. both $\vec{x}^{*}$ and $\vec{x}^{\#}$ are feasible, and $f(\vec{x}^{*}) < f(\vec{x}^{\#})$,
3. both $\vec{x}^{*}$ and $\vec{x}^{\#}$ are infeasible, and $\vec{x}^{*}$ dominates $\vec{x}^{\#}$.

More precisely, $\vec{x}^{*}$ replaces $\vec{x}^{\#}$ on condition that

$$
\begin{cases}
\forall i \in \{1, 2, \ldots, M\} : g_i(\vec{x}^*) \leq 0 \\
\wedge \\
\exists i \in \{1, 2, \ldots, M\} : g_i(\vec{x}^\#) > 0
\end{cases}
$$

$$\vee$$

$$
\begin{cases}
\forall i \in \{1, 2, \ldots, M\} : g_i(\vec{x}^*) \leq 0 \\
\wedge \\
\forall i \in \{1, 2, \ldots, M\} : g_i(\vec{x}^\#) \leq 0 \\
\wedge \\
f(\vec{x}^*) < f(\vec{x}^\#)
\end{cases}
$$

$$\vee$$

$$
\begin{cases}
\exists i \in \{1, 2, \ldots, M\} : g_i(\vec{x}^*) > 0 \\
\wedge \\
\exists i \in \{1, 2, \ldots, M\} : g_i(\vec{x}^\#) > 0 \\
\wedge \\
\forall i \in \{1, 2, \ldots, M\} : g_i(\vec{x}^*) \leq g_i(\vec{x}^\#) \\
\wedge \\
\exists i \in \{1, 2, \ldots, M\} : g_i(\vec{x}^*) < g_i(\vec{x}^\#).
\end{cases}
$$

Figure 4.3 illustrates these procedures of comparison between $\vec{x}^\#$ and $\vec{x}^*$, replacement of $\vec{x}^\#$ with $\vec{x}^*$, and elimination of $\vec{x}^*$.

It is observed from the above descriptions that the infeasible HM members violating the given constraints can also evolve in the modified HS method. In other words, we do not have to always search for new feasible HM members by repeatedly examining them with the constraint functions, as shown in Fig. 4.1. Compared with the original HS method, our approach needs only a considerably smaller number of constraint functions evaluations. However, we must emphasize that ranking of the HM members according to their objective and constraint functions values may involve a moderately higher computational complexity. More detailed performance evaluation and comparison of our modified HS method for constrained optimization can be found in Refs. [17, 18].

References

1. X.Z. Gao, X. Wang, S.J. Ovaska, Uni-modal and multi-modal optimization using modified harmony search methods. Int. J. Innov. Comput. Inf. Control. **5**(10a), 2985–2996 (2009)
2. A.M. Turkya, S. Abdullaha, A multi-population harmony search algorithm with external archive for dynamic optimization problems. Inf. Sci. **272**(10), 84–95 (2014)

3. S.O. Degertekin, Improved harmony search algorithms for sizing optimization of truss structures. Comput. Struct. **92–93**, 229–241 (2012)
4. M. Mahdavi, M. Fesanghary, E. Damangir, An improved harmony search algorithm for solving optimization problems. Appl. Math. Comput. **188**(2), 1567–1579 (2007)
5. M. Fesangharya, E. Damangira, I. Soleimanib, Design optimization of shell and tube heat exchangers using global sensitivity analysis and harmony search algorithm. Appl. Therm. Eng. **29**(5–6), 1026–1031 (2009)
6. M. El-Abd, An improved global-best harmony search algorithm. Appl. Math. Comput. **222**, 94–106 (2013)
7. V. Kumar, J.K. Chhabra, D. Kumar, Parameter adaptive harmony search algorithm for unimodal and multimodal optimization problems. J. Comput. Sci. **5**(2), 144–155 (2014)
8. M.G.H. Omrana, M. Mahdavib, Global-best harmony search. Appl. Math. Comput. **198**(2), 643–656 (2008)
9. F. Amini, P. Ghaderi, Hybridization of harmony search and ant colony optimization for optimal locating of structural dampers. Appl. Soft Comput. **13**(5), 2272–2280 (2013)
10. Q.K. Pan, L. Wang, L. Gao, A chaotic harmony search algorithm for the flow shop scheduling problem with limited buffers. Appl.Soft Comput. **11**(8), 5270–5280 (2011)
11. L.J. Li, Z.B. Huang, F. Liu et al., A heuristic particle swarm optimizer for optimization of pin connected structures. Comput. Struct. **85**(7–8), 340–349 (2007)
12. D. Zou, L. Gao, J. Wu, A novel global harmony search algorithm for reliability problems. Comput. Ind. Eng. **28**(2), 307–316 (2010)
13. S.H. Huang, P.C. Lin, A harmony-genetic based heuristic approach toward economic dispatching combined heat and power. Int. J. Electr. Power Energy Syst. **53**, 482–487 (2013)
14. A. Askarzadeh, A discrete chaotic harmony search-based simulated annealing algorithm for optimum design of PV/wind hybrid system. Sol. Energy **97**, 93–101 (2013)
15. B. Wu, C. Qian, W. Ni et al., Hybrid harmony search and artificial bee colony algorithm for global optimization problems. Comput. Math. Appl. **64**(8), 2621–2634 (2012)
16. E. Yazdi, V. Azizi, A.T. Haghighat, A new biped locomotion involving arms swing based on neural network with harmony search optimizer. in *IEEE International Conference on Automation and Logistics*, Chongqing, China, 15–16 June 2011
17. X.Z. Gao, X. Wang, S.J. Ovaska, A modified harmony search method in constrained optimization. Int. J. Innov. Comput. Inf. Control **6**(9), 4235–4247 (2009)
18. X.Z. Gao, X. Wang, S.J. Ovaska, Harmony search methods for multi-modal and constrained optimization, in *Music-Inspired Harmony Search Algorithm*, ed. by Z.W. Geem (Springer, Berlin, 2009), pp. 39–52
19. Z. Michalewicz, *Genetic algorithms + data structures = evolution programs*, 3rd edn. (Springer, Berlin, 1996)
20. C.A.C. Coello, Constraint-handling in genetic algorithms through the use of dominance-based tournament selection. Adv. Eng. Inf. **16**(3), 193–203 (2002)
21. K. Deb, An efficient constraint handling method for genetic algorithms. Comput. Methods Appl. Mech. Eng. **186**(2–4), 311–338 (2000)
22. X. Zhang, Q. Lu, S. Wen et al., A modified differential evolution for constrained optimization. ICIC Express Lett. **2**(2), 181–186 (2008)
23. C.A.C. Coello, Use of a self-adaptive penalty approach for engineering optimization problems. Comput. Ind. **41**(2), 113–127 (2000)
24. X. Wang, X. Yan, Global best harmony search algorithm with control parameters co-evolution based on PSO and its application to constrained optimal problems. Appl. Math. Comput. **219**(19), 10059–10072 (2013)
25. N. Sinsuphan, U. Leeton, T. Kulworawanichpong, Optimal power flow solution using improved harmony search method. Appl. Soft Comput. **13**(5), 2364–2374 (2013)
26. S. Kulluk, L. Ozbakir, A. Baykasoglu, Training neural networks with harmony search algorithms for classification problems. Eng. Appl. Artif. Intel. **25**(1), 11–19 (2012)

27. W K. Wong, Z.X. Guo, A hybrid intelligent model for medium-term sales forecasting in fashion retail supply chains using extreme learning machine and harmony search algorithm. In. J. Prod. Econ. **128**(2), 614–624 (2010)
28. M Mirkhani, R. Forsati, A.M. Shahri, A. Moayedikia, A novel efficient algorithm for mobile robot localization. Robot Auton. Syst. **61**(9), 920–931 (2013)

Chapter 5
The Hybrid Strategies of Harmony Search in Optimization Problem Solving

Abstract The remarkable capability of overcoming the shortcomings of NIC individual algorithms without losing their advantages makes the hybrid NIC techniques superior to the stand-alone ones and become a dominant research topic in the field of NIC methods. This hybridization mechanism has been demonstrated to achieve great improvement in the performance of different optimization issues. In this chapter, four hybrid optimization methods based on the HS combined with niching technique, opposition-based learning (OBL), clonal selection algorithm, and cultural algorithm are presented. The simulation results show that they can considerably outperform the original HS method in the applications of multimodal optimization, optimization of high-dimensional benchmark functions, minimization of weight of a tension/compression spring, optimal Sugeno fuzzy classification systems, and optimal design of welded beam, gear train, pressure vessel, and permanent magnet direct-driven wind generator.

Keywords Harmony search (HS) method · Hybrid harmony search (HS) methods · Multimodal optimization · Opposition-based learning · Cultural algorithm · Niching · Wind generator design

5.1 Niching HS Method for Multimodal Optimization

In this section, a niching HS algorithm n-HS is proposed, in which one of the most popular niching approaches, deterministic crowding (DC), is employed to modify the crossover operation used in the HS. The n-HS is compared with the original HS method using computer experiments on the optimization of a total of ten multimodal functions [1].

5.1.1 Niching Technique

The niching technique, such as fitness sharing, is one of the most popular approaches used in the multimodal optimization [2]. The fitness sharing is based

© The Author(s) 2015
X. Wang et al., *An Introduction to Harmony Search Optimization Method*,
SpringerBriefs in Computational Intelligence,
DOI 10.1007/978-3-319-08356-8_5

on the idea of scaling down the fitness of a chromosome in the population by the number of the other similar ones. However, the functions and parameters used in the fitness sharing including the similarity measure and sharing radius can significantly affect its performance. On the other hand, the tuning of the fitness-sharing technique usually requires some prior knowledge of the optimization problems, e.g., distances among the existing peaks in the search space.

The DC proposed by Mahfoud [3] is an alternative but efficient method for the multimodal optimization. In the DC, the crossover operator in the evolutionary computation creates offspring from the combination of selected parents. The offspring generated in this way may replace their closest parents, if they have higher fitness. The DC can be explained in more details in the following steps.

1. Randomly choose two different parents, p_1 and p_2, from the current population.
2. Produce two offspring, c_1 and c_2, with the crossover operation from p_1 and p_2.
3. Calculate the fitness of the offspring and their parents: $f(c_1), f(c_2), f(p_1)$, and $f(p_2)$. $f(\cdot)$ represents the fitness of these chromosomes.
4. Calculate the distances among the offspring and their parents: $d(p_1, c_1)$, $d(p_2, c_2)$, $d(p_1, c_2)$, and $d(p_2, c_1)$. $d(\cdot, \cdot)$ is the distance metric between two chromosomes.
5. The offspring can replace their parents according to the following rules:

IF $d(p_1, c_1) + d(p_2, c_2) > d(p_1, c_2) + d(p_2, c_1)$
IF $f(c_1) > f(p_1)$ THEN replace p_1 with c_1
IF $f(c_2) > f(p_2)$ THEN replace p_2 with c_2
ELSE
IF $f(c_2) > f(p_1)$ THEN replace p_1 with c_2
IF $f(c_1) > f(p_2)$ THEN replace p_2 with c_1
END

Apparently, in the DC method, the diversity of the offspring is effectively maintained on the basis of their fitness as well as the similarity between their parents and themselves. Nevertheless, it can be concluded from the above explanations of the HS, the technique that generates the new solution candidates from the HM members is analogous to the crossover operator used in the evolutionary computation. Therefore, the DC technique can be naturally generalized and embedded into the HS method so that it is capable of handling multimodal optimization problems. Inspired by this idea, we propose and study a niching HS, n-HS, in the next section.

5.1.2 A Niching HS Method for Multimodal Optimization

There have been quite many revised HS methods already proposed during the past decade to manipulate with various kinds of optimization problems, such as multiobjective optimization, dynamical optimization, and discrete optimization problems [4–6]. In Ref. [7], the idea of artificial fish algorithm is combined with

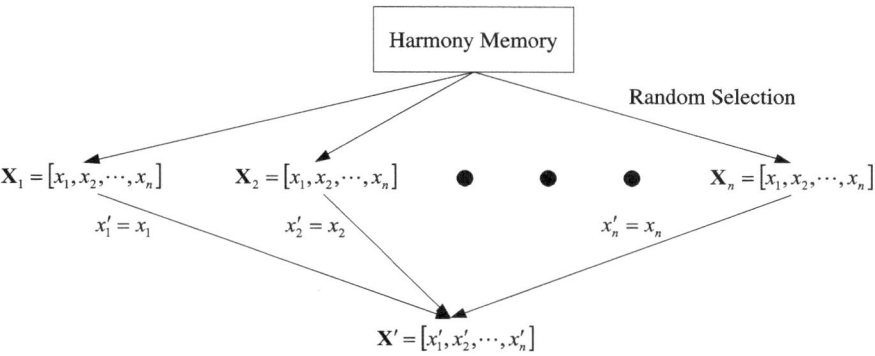

Fig. 5.1 Generation of new solution candidate $\mathbf{x}'$ using multiple harmony memory members in n-HS

the HS method by Gao et al. for the multimodal optimization problems. An effective diversity maintenance policy is employed for the members of the HS memory, in which the qualification of a solution candidate as a new harmony memory member should be based on not only its fitness but also its similarity to all the existing members. Unfortunately, this modified HS method may not yield satisfactory performances in case of poorly tuned parameters.

In the HS method, the new solution candidates can be created based on the multimember stored in the harmony memory. Thus, in order to utilize the DC technique, the similarity between this mechanism and the crossover operator used in the evolutionary computation should be built up. Step 2 of the original HS method is modified in our n-HS as follows. To generate a new solution candidate, $\mathbf{x}' = \left[x'_1, x'_2, \ldots, x'_n\right]$, a total of n members are first randomly selected from the harmony memory, which are denoted as $\mathbf{x}_1, \mathbf{x}_2, \ldots, \mathbf{x}_n$. It should be stressed that $\mathbf{x}_1, \mathbf{x}_2, \ldots, \mathbf{x}_n$ are usually mutually different. $\mathbf{x}'$ is then constructed in the following way. The first component of $\mathbf{x}'$ comes from the first component of $\mathbf{x}_1$, the second component from the second component of $\mathbf{x}_2$, etc. More details of the generation of $\mathbf{x}'$ are illustrated in Fig. 5.1.

Based on the principle of the DC, the distances between $\mathbf{x}'$ and $\mathbf{x}_1, \mathbf{x}_2, \ldots, \mathbf{x}_n$ are calculated as $d_1, d_2, \ldots, d_n$. The harmony member $\mathbf{x}_i$ that has the minimal distance to $\mathbf{x}'$ is selected. If the fitness of $\mathbf{x}'$ is greater than that of $\mathbf{x}_i$, the latter will be replaced by the former. Otherwise, $\mathbf{x}'$ is disregarded, and the iteration procedure continues.

IF $f(\mathbf{x}') > f(\mathbf{x}_i)$
replace $\mathbf{x}_i$ with $\mathbf{x}'$
ELSE
disregard $\mathbf{x}'$
END

Note that the above DC-like policy only applies to the solution candidates generated from the harmony memory members. In other words, according to HMCR, if $\mathbf{x}'$ is randomly created, it will still follow Step 3 in the regular HS method.

It is obvious that the DC approach used in the n-HS can prevent the harmful over-similarity among the harmony memory members so that the diversity of the HS solutions can be effectively maintained. That is to say, this modified HS method is well suited for handling the multimodal problems. Nevertheless, the n-HS has two drawbacks. Firstly, since $\mathbf{x}'$ is compared with the multiparent from the harmony memory instead of only the worst harmony memory member, it may take a relatively long iteration time for the n-HS to converge. Secondly, the distances between $\mathbf{x}'$ and $\mathbf{x}_1, \mathbf{x}_2, \ldots, \mathbf{x}_n, d_1, d_2, \ldots, d_n$, have to be calculated. This requirement can certainly result in a time-consuming procedure in case of a very large value of n. The next section demonstrates and verifies the multimodal optimization capability of our n-HS on the basis of ten multimodal functions.

5.1.3 Simulations

The multimodal optimization performance of the proposed n-HS method is examined using the following ten two-dimensional functions [8, 9]. Each function may have several global as well as local optima, as shown in Fig. 5.2a–j. Actually, the goal of the optimization algorithms employed here is to find as many of them as possible.

1. F1 function (Branin function):

$$f(x, y) = \left(y - \frac{5.1}{4\pi^2}x^2 + \frac{5}{\pi}x - 6\right)^2 + 10\left(1 - \frac{1}{8\pi}\right)\cos(x) + 10, \tag{5.1}$$
$$x, y \in [-5, 15].$$

There are three global minimum solutions at $[-\pi, 12.275]$, $[\pi, 2.275]$, and $[9.42478, 2.475]$, respectively, which lead to the minimal function output $f^*(x, y) = 0.397887$.

2. F2 function (Four-peak function):

$$f(x) = e^{\left[-(x-4)^2-(y-4)^2\right]} + e^{\left[-(x+4)^2-(y-4)^2\right]}$$
$$+ 2\left\{e^{\left(-x^2-y^2\right)} + e^{\left[-x^2-(y+4)^2\right]}\right\}, \quad x, y \in [-5, 5]. \tag{5.2}$$

There are two global maximal solutions at $[0, 0]$ and $[0, -4]$, and two local maxima at $[-4, 4]$ and $[4, 4]$, respectively. The maximal function output is $f^*(x, y) = 2$.

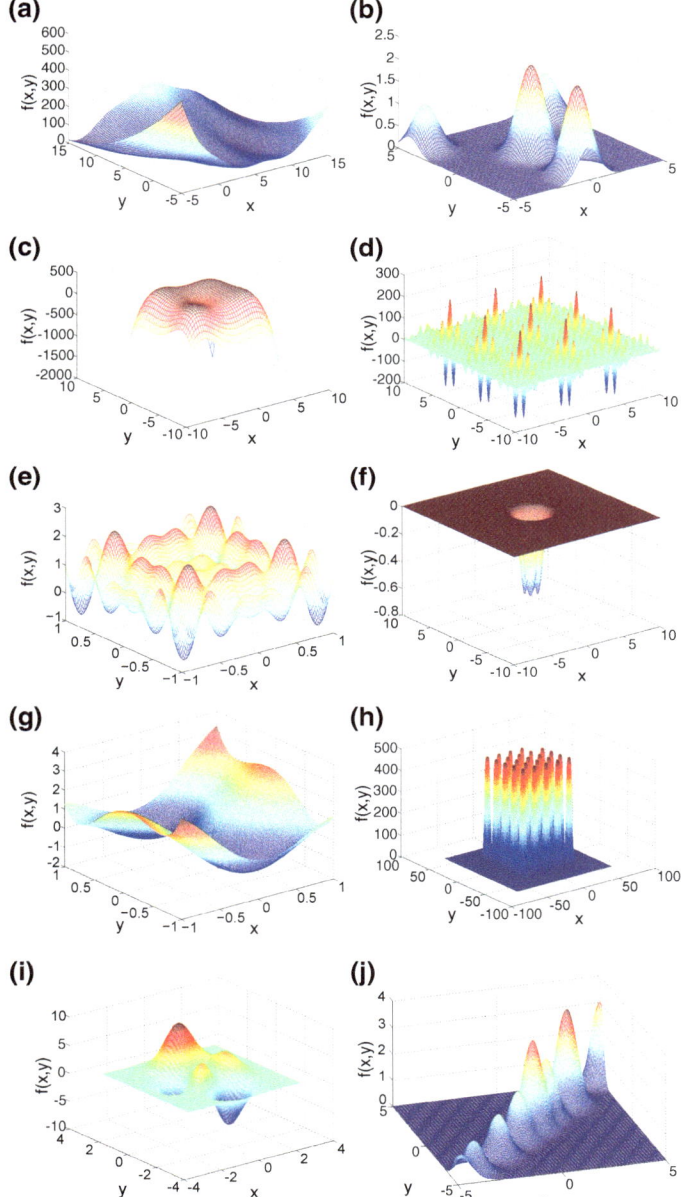

Fig. 5.2 Ten multimodal functions used in simulations

3.   F3 function (Himmelblau function)

$$f(x,y) = 200 - \left(x^2 + y - 11\right)^2 + \left(x + y^2 - 7\right)^2, \quad -5 \le x, y \le 5. \tag{5.3}$$

There are four global maximal solutions at, $[3.58, -1.86]$ $[3, 2]$, $[-2.815, 3.125]$, and $[-3.78, -3.28]$, respectively. The maximal function output is $f^*(x, y) = 200$.

4.   F4 function (Shubert function)

$$f(x,y) = \sum_{i=1}^{5} i \cos[i + (i+1)x] \sum_{i=1}^{5} i \cos[i + (i+1)y], \quad -10 \le x, y \le 10. \tag{5.4}$$

There are a total of 18 global maximal solutions. The maximal function output is $f^*(x, y) = -186.7$.

5.   F5 function

$$f(x,y) = x \sin(4\pi x) - y \sin(4\pi y + \pi) + 1, \quad -1 \le x, y \le 1. \tag{5.5}$$

There are four global maximal solutions at $[0.6349, 0.6349]$, $[-0.6349, 0.6349]$, $[0.6349, -0.6349]$, and $[-0.6349, -0.6349]$, respectively. The maximal function output is $f^*(x, y) = 2.26$.

6.   F6 function (Yang function)

$$f(x,y) = -(|x| + |y|)e^{-\left(x^2 + y^2\right)}, \quad -10 \le x, y \le 10. \tag{5.6}$$

There are four global minimal solutions at $[0.5, 0.5]$, $[-0.5, 0.5]$, $[0.5, -0.5]$, and $[-0.5, -0.5]$, respectively. The minimal function output is $f^*(x, y) = -0.6065$.

7.   F7 function (Six-hump camel back function)

$$f(x,y) = \left(4 - 2.1x^2 + \frac{1}{3}x^4\right)x^2 + xy + 4\left(y^2 - 1\right)y^2, \quad -1 \le x, y \le 1. \tag{5.7}$$

There are two global minimal solutions at $[0.0898, -0.7126]$ and $[-0.0898, 0.7126]$, respectively. The minimal function output is $f^*(x, y) = -1.0316$.

8.   F8 function (Shekel's Foxholes function)

$$f(x,y) = \frac{1}{0.002 + \sum_{i=0}^{24} \frac{1}{i + \left(x - a_{1j}\right)^6 + \left(y - a_{2j}\right)^6}}, \quad -65.536 \le x, y \le 65.536. \tag{5.8}$$

where $a_{ij} = \begin{bmatrix} -32 & -16 & 0 & 16 & 32 & \cdots & 0 & 16 & 32 \\ -32 & -32 & -32 & -32 & -32 & \cdots & 32 & 32 & 32 \end{bmatrix}$.

There are a total of 25 global and local maximal solutions.

9.   F9 function (Peaks function)

$$f(x,y) = 3(1-x)^2 e^{-[x^2+(y+1)^2]} - 10\left(\frac{x}{5} - x^3 - y^5\right)e^{-(x^2+y^2)}$$
$$-\frac{1}{3}e^{-[(x+1)^2+y^2]}, \quad -3 \le x, y \le 3. \tag{5.9}$$

There are three maximal solutions at [0, 1.58], [−0.46, −0.63], and [1.28, 0], respectively.

10.   F10 function (Random-Peaks function)

$$f(x,y) = \sum_{i=1}^{Q} a_i e^{\left\{-b_i\left[(x-x_i)^2+(y-y_i)^2\right]\right\}}, \quad -5 \le x, y \le 5. \tag{5.10}$$

Q is the number of peaks, and $(x_i - y_i)$ is the location of each peak. a_i, b_i, x_i, y_i are given as follows:

$$a_i = 1 + 2v, \tag{5.11}$$

$$b_i = 2 + v, \tag{5.12}$$

$$x_i = y_i = -5 + 10v, \tag{5.13}$$

where v is uniformly distributed within the interval [0, 1]. In our simulations, $Q = 10$, which means that there are ten maximal solutions to be located.

Both the HS and n-HS are employed to obtain the optimal solutions to these functions, and their performances are compared with each other. There are 500 members in the harmony memory of these two HS methods, i.e., HMS = 500. The other parameters used in the HS and n-HS are given as follows: HMCR = 0.85 and PAR = 0.35. It should be pointed out that all the optimization results presented here are based on the average of 1,000 independent trials. To measure the numbers of the optima located, a solution acquired is considered to be an optimum, if the distance between these two solutions is smaller than a preset radius threshold. These radius thresholds for the multimodal functions together with the optimal solutions found by the HS and n-HS are given in Table 5.1. As two illustrative examples, the optimization results of F3 and F5 functions are demonstrated in Figs. 5.3 and 5.4, respectively. Apparently, the n-HS is capable of efficiently locating most of the global and local optima, while the regular HS method is always trapped into only one or two of them. Therefore, to summarize, the DC technique embedded in our n-HS can indeed result in a superior multimodal optimization performance.

In summary, simulation examples of ten multimodal functions are used to verify the efficiency of the proposed n-HS. It has been demonstrated to yield a significantly better multimodal optimization performance than that of the regular HS method.

Table 5.1 Optimal solutions to multimodal functions located by HS and n-HS

	Number of optima	Radius threshold	Number of optima obtained by HS	Number of optima obtained by n-HS
F1 function	3	0.1	1.0840	2.6840
F2 function	4	0.01	1.0030	3.9140
F3 function	4	0.01	1.1990	3.9080
F4 function	18	0.01	1.6870	17.3720
F5 function	4	0.01	1.1060	3.6440
F6 function	4	0.01	1.0260	3.8610
F7 function	2	0.01	1.0550	2.0000
F8 function	25	0.25	1.8500	22.6110
F9 function	3	0.01	0.9980	2.9810
F10 function	10	0.01	0.1310	8.5870

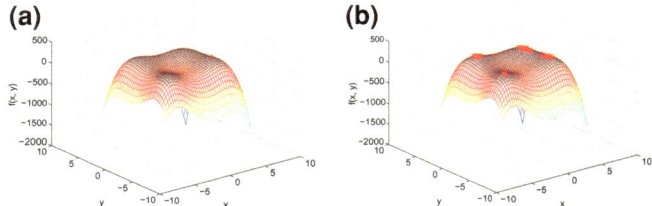

Fig. 5.3 Multiple optima of F3 function obtained by HS and n-HS. **a** Multiple optima obtained by HS and **b** multiple optima obtained by n-HS

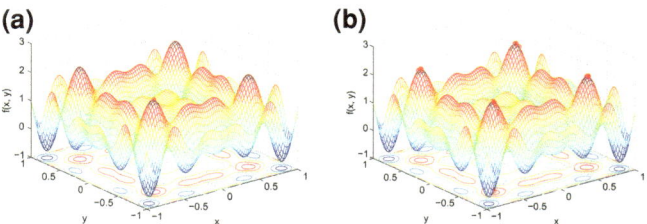

Fig. 5.4 Multiple optima of F5 function obtained by HS and n-HS. **a** Multiple optima obtained by HS and **b** multiple optima obtained by n-HS

5.2 A Hybrid Optimization Method of HS and Opposition-Based Learning

In this section, a hybridization of the HS and opposition-based learning (OBL) (HS-OBL) is developed and explored. In the HS-OBL, the OBL is employed to enhance the mutation operation of the HS, which can lead to a better convergence performance [5].

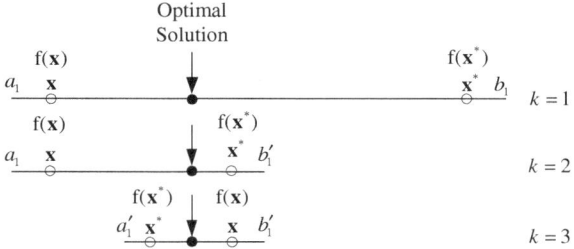

Fig. 5.5 Opposition-based learning (*OBL*) in one-dimensional ($n = 1$) optimization

5.2.1 Opposition-Based Learning

Developed by Tizhoosh, the OBL is a new approach to machine intelligence, which has been widely employed in optimization, neural networks training, and reinforcement learning [10]. For dealing with optimization problems, the OBL is based on the utilization of the opposition numbers of the current search directions. More precisely, suppose $\mathbf{x} = (x_1, x_2, \ldots, x_n)$ is a single search point in the n-dimensional solution space, and $x_i \in [a_i, b_i]$, $i = 1, 2, \ldots, n$. To simplify our presentation, only the continuous variables $\mathbf{x}$ are considered here. The opposition number $\mathbf{x}^* = \left(x_1^*, x_2^*, \ldots, x_n^*\right)$ of $\mathbf{x} = (x_1, x_2, \ldots, x_n)$ is defined as follows:

$$x_i^* = a_i + b_i - x_i, \quad i = 1, 2, \ldots, n. \tag{5.14}$$

The principle of the OBL for optimization is that the search for the optimal solutions should be on the basis of both $\mathbf{x}$ and $\mathbf{x}^*$ as follows:

In every iteration, $\mathbf{x}^*$ is calculated from $\mathbf{x}$, and let $f(\mathbf{x})$ and $f(\mathbf{x}^*)$ represent the fitness of $\mathbf{x}$ and $\mathbf{x}^*$, respectively. The iterations proceed with $\mathbf{x}$, if $f(\mathbf{x}) \geq f(\mathbf{x}^*)$, otherwise, with $\mathbf{x}^*$. Note that '$\geq$' here means 'better than or equal to with regard to the objective function $f(\mathbf{x})$.' An illustrative example of the OBL in the simple one-dimensional ($n = 1$) optimization case is given in Fig. 5.5, where k is the iteration step.

As Fig. 5.5 shows, with the growth of OBL iterations, the search interval can be *recursively* shrunk by half by choosing the solution candidate as the better one between $\mathbf{x}$ and $\mathbf{x}^*$. This procedure will ultimately converge, when $\mathbf{x}$ has approached to be close enough with $\mathbf{x}^*$. From these descriptions, it is concluded that the counterpart of $\mathbf{x}$ is utilized in the OBL so that the efficiency of search can be improved. Particularly, the employment of the OBL in the GA, reinforcement learning, and DE has been discussed in [11–13], respectively. In the next section, a hybrid HS, HS-OBL, is proposed and studied by incorporating the OBL into the original HS method.

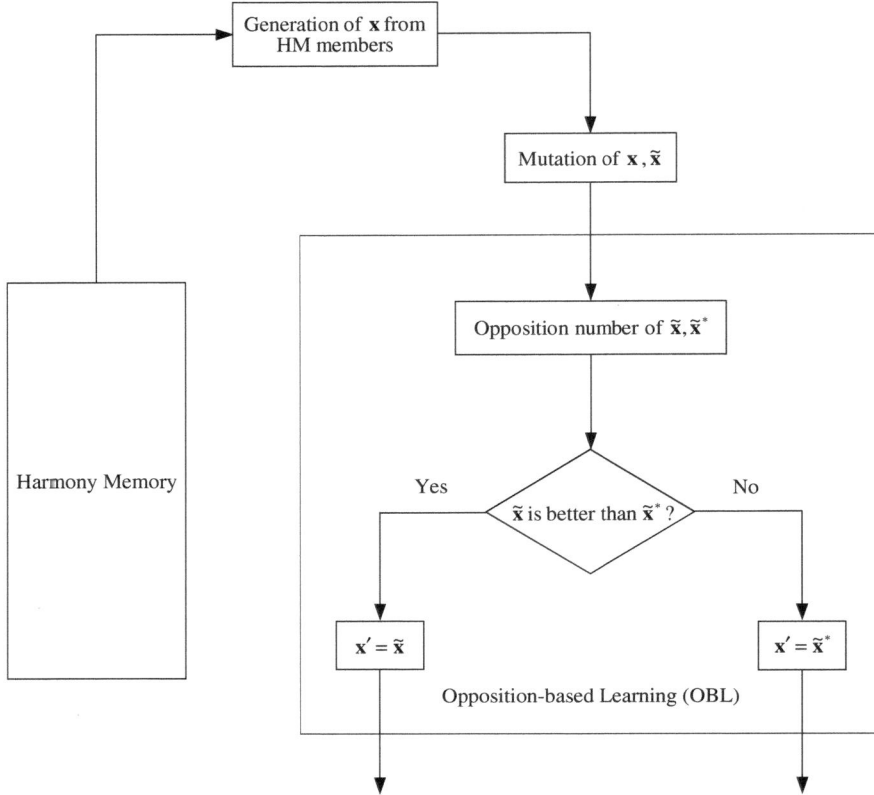

Fig. 5.6 Hybridization of HS and OBL: HS-OBL

5.2.2 Fusion of HS and OBL: HS-OBL

In this section, the OBL discussed above is embedded into the HS to acquire an enhanced optimization performance, and a hybrid HS method, HS-OBL, is proposed accordingly. As aforementioned, the new HM members in the HS method can be obtained in two basic ways: combination/mutation of the existing HM members and random generation. Particularly, the HS-OBL employs the OBL in the mutation operation of the HS method. The diagram of our HS-OBL is illustrated in Fig. 5.6.

To put it into more details, let x denote the new solution candidate that is generated from the combination of the existing HM members. Subject to the HS parameter PAR, the mutated x, $\tilde{x}$, is first calculated. Next, based on the predefined OBL probability [10], P^{OBL}, the opposition number of $\tilde{x}$, $\tilde{x}^*$, can be also obtained by using the OBL strategy. In other words, the OBL operation is subject to P^{OBL} in the HS-OBL. Note that $[a_i, b_i]$, $i = 1, 2, \ldots, n$ are determined by the ranges of the current HM members. The fitness of both $\tilde{\mathbf{x}}$ and $\tilde{\mathbf{x}}^*$ is evaluated and is then compared with each other. Only the one with the better fitness will be retained as the new solution candidate for the HM to evolve in the next iteration. It is also emphasized that the ranges of $[a_i, b_i]$ shrink with the growth of iterations of the HS-OBL, which may result in the enhanced convergence of the hybrid optimization method. That is to say, the search ranges should be halved recursively based on which of $\tilde{\mathbf{x}}$ and $\tilde{\mathbf{x}}^*$ is better with regard to their fitness. This iteration procedure is repeated until a preset termination criterion is satisfied. The pseudo-codes of the HS-OBL are given in Fig. 5.7.

It can be concluded that the deployment of the OBL in the HS-OBL is indeed beneficial for creating high-quality solution candidates and, therefore, enhances the overall convergence characteristics of the original HS method. Unfortunately, the generation and evaluation of $\tilde{\mathbf{x}}^*$ might increase the computational complexity of the HS method as well. In the following section, the performances of the HS and HS-OBL are compared using numerical simulations of the optimization of a total of 24 representative nonlinear functions and a tension/compression spring.

5.2.3 Optimization of Nonlinear Functions and a Tension/ Compression Spring Using HS-OBL

In this section, the performance comparison is investigated between the original HS and HS-OBL with the simulation examples of some nonlinear functions and an optimal spring design problem. The functions are well-known benchmarks for testing modern optimization methods, and they all belong to the minimization problems. A total of 24 popular test-bed functions are considered here, which can be classified into two main categories: low-dimensional and high-dimensional functions.

5.2.3.1   Low-dimensional Functions

The following 11 low-dimensional functions [8] are first examined in the simulations.

Aluffi–Pentini function:

$$f(x) = 0.25x_1^4 - 0.5x_1^2 + 0.1x_1 + 0.5x_2^2, \quad x \in [-10, 20]. \tag{5.15}$$

Beale function:

$$f(x) = [1.5 - x_1(1 - x_2)]^2 + [2.25 - x_1(1 - x_2^2)]^2 \\ + [2.625 - x_1(1 - x_2^3)]^2, \quad x \in [-4.5, 9]. \tag{5.16}$$

/* HM initialization */

for (i=1; i<= HMS; i++)

   for (j=1; j<=n; j++)

      Randomly initialize x_j^i in HM.

   endfor

endfor

/* End of HM initialization */

Repeat

   /* New solution candidates construction and evaluation */

   if (rand(0, 1)<HMCR)

      Construct solution candidate $\mathbf{X}$ based on combination of the current HM members.

      if (rand(0, 1)<PAR)

         Apply pitch adjustment distance bw to mutate $\mathbf{X}$:

         $\tilde{\mathbf{x}} = \mathbf{x} \pm \text{rand}(0,1) \times bw$.

      endif

      Evaluate the fitness of $\tilde{\mathbf{x}}$: $f(\tilde{\mathbf{x}})$.

      /* Opposition number $\tilde{\mathbf{x}}^*$ construction and evaluation */

      if (rand(0, 1)< P^{OBL})

         Generate opposition number $\tilde{\mathbf{x}}^*$ from $\tilde{\mathbf{x}}$ using (2).

         Evaluate the fitness of $\tilde{\mathbf{x}}^*$: $f(\tilde{\mathbf{x}}^*)$.

         if ($f(\tilde{\mathbf{x}}^*)$ is better than $f(\tilde{\mathbf{x}})$)

            $\mathbf{x} = \tilde{\mathbf{x}}^*$.

         else

            $\mathbf{x} = \tilde{\mathbf{x}}$.

         endif

      endif

      /* End of opposition number $\mathbf{x}^*$ construction and evaluation */

   else

      Construct solution candidate $\mathbf{X}$ randomly.

      Evaluate the fitness of $\mathbf{X}$: $f(\mathbf{x})$.

   endif

   /* End of new solution candidates construction and evaluation */

   /* HM update */

   if ($f(\mathbf{x})$ is better than the fitness of the worst HM member)

      Replace the worst HM member with $\mathbf{X}$.

   endif

   /* End of HM update */

Until a preset termination criterion is met.

Fig. 5.7 Pseudo-codes of HS-OBL

Becker and Lago function:

$$f(x) = (|x_1| - 5)^2 + (|x_2| - 5)^2, \quad x \in [-10, 20].   \tag{5.17}$$

Bohachevsky function:

$$f(x) = x_1^2 + 2x_2^2 - 0.3\cos(3\pi x_1) - 0.4\cos(4\pi x_2) + 0.7, \quad x \in [-50, 100].   \tag{5.18}$$

Colville function:

$$\begin{aligned}
f(x) = {}& 100(x_2 - x_1^2)^2 + (1 - x_1)^2 + 90(x_4 - x_3^2)^2 \\
& + (1 - x_3)^2 + 10.1\left[(x_2 - 1)^2 + (x_4 - 1)^2\right] \\
& + 19.8(x_2 - 1)(x_4 - 1), \quad x \in [-10, 20].
\end{aligned}   \tag{5.19}$$

Easom function:

$$f(x) = -\cos(x_1)\cos(x_2)e^{-(x_1-\pi)^2-(x_2-\pi)^2}, \quad x \in [-100, 200].   \tag{5.20}$$

Goldstein and Price function:

$$\begin{aligned}
f(x) = {}& \left[1 + (x_1 + x_2 + 1)^2(19 - 14x_1 + 3x_1^2 - 14x_2 + 6x_1x_2 + 3x_2^2)\right] \\
& \times \left[30 + (2x_1 - 3x_2)^2(18 - 32x_1 + 12x_2^2 + 48x_2 - 36x_1x_2 + 72x_2^2)\right], \\
& x \in [-2, 4].
\end{aligned}   \tag{5.21}$$

Hosaki function:

$$f(x) = \left(1 - 8x_1 + 7x_1^2 - \frac{7}{3}x_1^3 + \frac{1}{4}x_1^4\right)x_2^2 e^{-x_2}, \quad x_1 \in [0, 5] \quad x_2 \in [0, 7].   \tag{5.22}$$

Periodic function:

$$f(x) = 1 + \sin(x_1)^2 + \sin(x_2)^2 - 0.1e^{(-x_1^2-x_2^2)}, \quad x \in [-10, 20].   \tag{5.23}$$

Table 5.2 Optimal solutions to low-dimensional functions acquired by HS and HS-OBL within 10,000 NFE

	Global minima	HS	HS-OBL
Aluffi–Pentini function	−0.3524	−0.3520	−0.35238
Beale function	0	1.5404×10^{-4}	1.2965×10^{-6}
Becker and Lago function	0	2.3161×10^{-8}	1.0870×10^{-9}
Bohachevsky function	0	3.9299×10^{-5}	2.3043×10^{-6}
Branin function	0.397887	0.3980	0.3979
Colville function	0	8.5924	0.3900
Easom function	−1	−0.9900	−0.999997
Goldstein and Price function	3	3.001	3.00000003
Hosaki function	−2.345811576	−2.345811554	−2.345811574
Periodic function	0.9	0.93	0.906
Wood function	0	9.34	0.39

Wood function:

$$
\begin{aligned}
f(x) = {} & 100\left(x_2 - x_1^2\right)^2 + (1 - x_1)^2 + 90\left(x_4 - x_3^2\right)^2 \\
& + (1 - x_3)^2 + 10.1\left[(x_2 - 1)^2 + (x_4 - 1)^2\right] \\
& + 19.8(x_2 - 1)(x_4 - 1), \quad x \in [-10, 20].
\end{aligned}
\tag{5.24}
$$

It should be stressed that different from the conventional function settings, the variable ranges of these functions used here are all extended and asymmetric, which may considerably increase the difficulty of the optimization tasks. Generally, evaluation of the objective function is the most time-consuming part in a lot of practical optimization systems. Thus, the number of function evaluation (NFE) rather than number of evolving iterations is used as the principal criterion to compare the convergence speeds of the HS and HS-OBL. Both of them have 100 HM members, i.e., HMS = 100, which are always initialized to be equal. The common parameters in these two optimization methods are given as follows: HMCR = 0.8 and PAR = 0.6. However, in the HS-OBL, the OBL coefficient, P^{OBL}, is chosen to be $P^{\mathrm{OBL}} = 0.35$. Their evolution procedures are terminated after 10,000 NFE. It is pointed out that the optimization results here are based on the average of 1,000 independent trials. All the simulations in this section are made under the MATLAB 7.0 environment on a Dell OptiPlex GX 745 computer with a 2.4-GHz Core 2 Duo E6600 CPU and 2-G system memory.

The optimal solutions to the above low-dimensional functions acquired by the HS and HS-OBL are provided in Table 5.2. Apparently, compared with the original HS method, for all the 10 low-dimensional functions, the HS-OBL can achieve moderately better optimization results within the same NFE, due to the deployment of the embedded OBL. That is, the HS-OBL has a superior nonlinear function optimization capability.

5.2.3.2 High-dimensional Functions

Next, the below 13 n-dimensional ($n = 10$ and $n = 50$) nonlinear functions [14] are also used in the performance comparison of the HS and HS-OBL.
Ackley function:

$$f(x) = -20e^{-0.2\sqrt{\frac{1}{n}\sum_{i=1}^{n} x_i^2}} - e^{\frac{1}{n}\sum_{i=1}^{n} \cos(2\pi x_i)} + 20 - e, \quad x \in [-32, 64]. \quad (5.25)$$

Alpine function:

$$f(x) = \sum_{i=1}^{n} |x_i \sin(x_i) + 0.1x_i|, \quad x \in [-10, 20]. \quad (5.26)$$

Bohachevsky function:

$$f(x) = \sum_{i=1}^{n} \left[x_i^2 + 2x_{i+1}^2 - 0.3\cos(3\pi x_i) - 0.4\cos(4\pi x_{i+1}) + 0.7 \right], \quad x \in [-15, 30]. \quad (5.27)$$

De Jong function:

$$f(x) = \sum_{i=1}^{n} i x_i^4, \quad x \in [-1.28, 2.56]. \quad (5.28)$$

Griewank function:

$$f(x) = \frac{1}{4,000} \sum_{i=1}^{n} x_i^2 - \prod_{i=1}^{n} \cos\frac{x_i}{\sqrt{i}} + 1, \quad x \in [-20, 10]. \quad (5.29)$$

Hyper-ellipsoid function:

$$f(x) = \sum_{i=1}^{n} 2^i x_i^2, \quad x \in [-100, 200]. \quad (5.30)$$

Levy function:

$$f(x) = \sin^2(3\pi x_1) + \sum_{i=1}^{n-1} (x_i - 1)^2 \left[1 + \sin^2(3\pi x_{i+1}) \right]$$
$$+ (x_n - 1)\left[1 + \sin^2(3\pi x_n) \right], \quad x \in [-20, 10]. \quad (5.31)$$

Powell function:

$$f(x) = \sum_{i=1}^{\frac{n}{4}} (x_{4i-3} + 10x_{4i-2})^2 + 5(x_{4i-1} - x_{4i})^2$$
$$+ (x_{4i-2} - x_{4i-1})^4 + 10(x_{4i-3} - x_{4i})^2, \quad x \in [-4, 5].$$

(5.32)

Restrigin function:

$$f(x) = \sum_{i=1}^{n} x_i^2 + 10 - 10\cos(2\pi x_i), \quad x \in [-5.12, 10.24]. \tag{5.33}$$

Rosenbrock function:

$$f(x) = \sum_{i=1}^{n} 100(x_{i+1} - x_i^2)^2 + (x_i - 1)^2, \quad x \in [-20, 10]. \tag{5.34}$$

Schaffer function:

$$f(x) = \sum_{i=1}^{n} (x_i^2 + x_{i+1}^2)^{0.25} \left\{ \left[\sin 50(x_i^2 + x_{i+1}^2)^{0.1} \right]^2 + 1 \right\}, \quad x \in [-100, 200].$$

(5.35)

Sphere function:

$$f(x) = \sum_{i=1}^{n} x_i^2, \quad x \in [-200, 100]. \tag{5.36}$$

Trid function:

$$f(x) = \sum_{i=1}^{n} (x_i - 1)^2 - \sum_{i=2}^{n} x_i x_{i-1}, \quad x \in [-2500, 5000]. \tag{5.37}$$

The global minima of all these functions are at $f(x) = 0$, except for the Trid function, whose global minimum is unknown when $n = 50$.

The same algorithm parameters as in the low-dimensional functions' case are used here. Again, the simulation results are on the basis of 1,000 separate trials. Tables 5.3 and 5.4 present the average optimal solutions to the above high-dimensional functions acquired by the HS and HS-OBL, when $n = 10$ and $n = 50$, respectively. As an illustrative example, Figs. 5.8 and 5.9 show the iteration procedures (average over 1,000 trials) of the HS and HS-OBL in optimizing the Powell function, when $n = 10$

Table 5.3 Average optimal solutions to high-dimensional functions acquired by HS and HS-OBL within 10,000 NFE over 1,000 trials ($n = 10$)

	HS	HS-OBL
Ackley function	10.1928	4.8698
Alpine function	0.0304	0.0228
Bohachevsky function	3.4590	2.3355
De Jong function	2.0111×10^{-5}	2.5953×10^{-9}
Griewank function	21.2909	21.2527
Hyper-ellipsoid function	525.2287	31.5621
Levy function	0.2841	0.0756
Powell function	0.8469	0.0097
Rastrigin function	8.1405	3.4346
Rosenbrock function	72.4170	13.5661
Schaffer function	10.2531	6.7364
Sphere function	0.2935	0.2854
Trid function	-128.8309	-164.4794

Table 5.4 Average optimal solutions to high-dimensional functions acquired by HS and HS-OBL within 10,000 NFE over 1,000 trials ($n = 50$)

	HS	HS-OBL
Ackley function	16.8903	16.5508
Alpine function	4.6604	4.5265
Bohachevsky function	240.8577	216.1322
De Jong function	1.7505	1.2189
Griewank function	103.2738	102.2558
Hyper-ellipsoid function	1.6328×10^{15}	2.4658×10^{14}
Levy function	11.0475	10.3347
Powell function	175.5891	113.1235
Rastrigin function	91.0487	87.9726
Rosenbrock function	1.4169×10^{4}	1.4064×10^{4}
Schaffer function	186.6432	184.4455
Sphere function	1.2273×10^{3}	1.1281×10^{3}
Trid function	2.4030×10^{6}	2.0473×10^{6}

and $n = 50$, respectively. It can be concluded that the HS-OBL is indeed well capable of outperforming the original HS method in the optimization of the selected low-dimensional and high-dimensional functions.

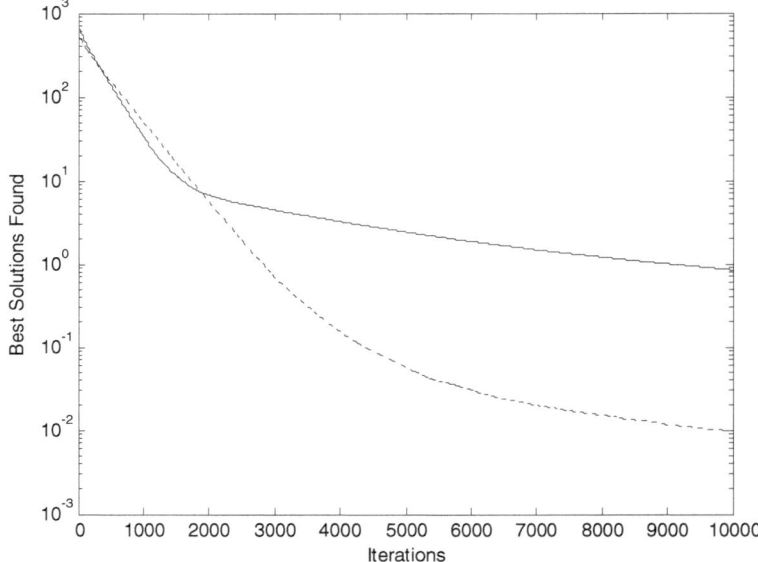

Fig. 5.8 Average optimization procedures of Powell function ($n = 10$) with HS (*solid line*) and HS-OBL (*dotted line*)

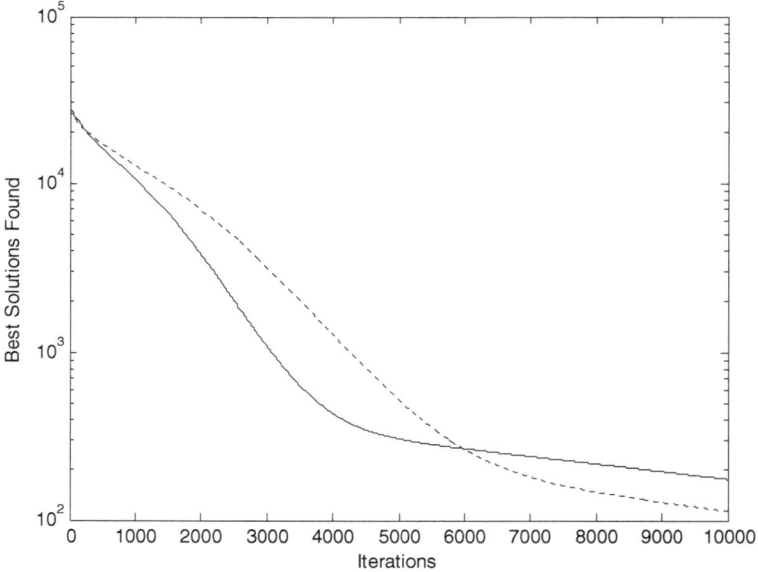

Fig. 5.9 Average optimization procedures of Powell function ($n = 50$) with HS (*solid line*) and HS-OBL (*dotted line*)

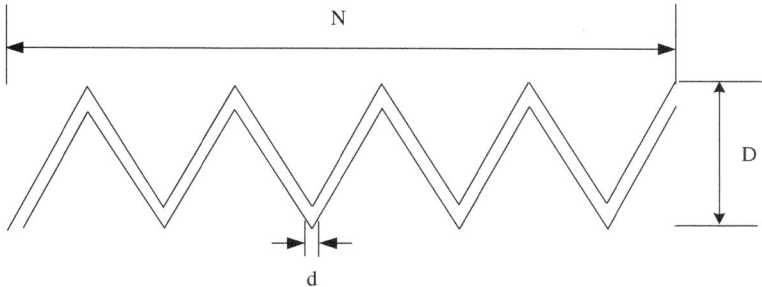

Fig. 5.10 Tension/compression spring

5.2.3.3 Minimization of Weight of a Tension/Compression Spring

This real-world optimization problem targets at minimizing the weight of a tension/compression spring subject to the constraints on its shear stress, surge frequency, and minimum deflection [15]. The parameters to be optimized are the mean coil diameter D, wire diameter d, and number of active coils N. A tension/compression spring with these design parameters is shown in Fig. 5.10. The spring design problem can be explained as follows:

$$\text{minimize} \quad f(\vec{x}) = x_1^2 x_2 (x_3 + 2),$$

$$\text{subject to} \quad g_1(\vec{x}) = 1 - \frac{x_2^3 x_3}{71,785 x_1^4} \leq 0,$$

$$g_2(\vec{x}) = \frac{4x_2^2 - x_1 x_2}{12,566 \left(x_1^3 x_2 - x_1^4\right)} + \frac{1}{5,108 x_1^2} - 1 \leq 0,$$

$$g_3(\vec{x}) = 1 - \frac{140.45 x_1}{x_2^2 x_3} \leq 0,$$

$$g_4(\vec{x}) = \frac{x_1 + x_2}{1.5} - 1 \leq 0,$$

where x_1 is D, x_2 is d, and x_3 is N. $0.05 \leq x_1 \leq 2$, $0.25 \leq x_2 \leq 1.3$, $2 \leq x_3 \leq 15$.

Both the HS and HS-OBL are employed here to deal with the above-optimal spring design problem. The two HS methods are configured with the same parameters as in the nonlinear function optimization case. Their optimization results are given in Table 5.5. Obviously, compared with the original HS, the HS-OBL can provide a better objective function value $f(\vec{x}) = 0.012665$.

Table 5.5 Average optimization results of weight of tension/compression spring using HS and HS-OBL

	HS	HS-OBL
x_1	0.050618	0.051677
x_2	0.331358	0.356416
x_3	12.963850	11.306742
$g_1(\vec{x})$	-8.748167×10^{-4}	-6.133399×10^{-6}
$g_2(\vec{x})$	-3.121165×10^{-4}	-9.816884×10^{-7}
$g_3(\vec{x})$	-3.994547	-4.053154
$g_4(\vec{x})$	-0.745349	-0.727938
$f(\vec{x})$	0.012704	0.012665

5.3 Fusion of Clonal Selection Algorithm and HS Method

In this section, the CSA is employed to improve the harmony memory members in the HS method. The developed hybrid optimization algorithm is further used to optimize the Sugeno fuzzy classification systems for the Fisher iris data and wine data classification [16].

5.3.1 Hybrid HS Optimization Algorithm

We develop a hybrid HS optimization technique based on the fusion of the CSA and HS method in this section. As aforementioned, the elite maintenance policy is a distinguishing property of the HS method and has a central effect on its behaviors. However, the update of the HS memory highly depends on the past search experiences. This inherent shortcoming limits the search ability of the regular HS method, especially in handling complex optimization problems. In our novel approach, the CSA is employed to improve the fitness of the solution candidates in the HM. That is to say, all the members of the HM are regarded as the individual antibodies, and they can evolve in the population of the CSA. For example, $[x_1^i, x_2^i, \ldots, x_n^i]$ is updated to $[x'_1^i, x'_2^i, \ldots, x'_n^i]$ so as to gain a better affinity with the antigen after a certain number of the CSA iterations. The CSA-based update of the HM members is indeed embedded into the HS method as a separate fine-tuning approach. Figure 5.11 illustrates how the CSA is merged with the HS method in our hybrid HS optimization scheme.

The proposed hybrid HS optimization algorithm takes the advantages from both the CSA and HS. The CSA-aided tuning strategy can provide a set of diverse members for the HM, which results in an improved convergence capability to deal with the premature problem. In addition, it should be stressed that the CSA only moderately increases the computational complexity of the original HS method.

Fig. 5.11 Hybrid
optimization method based
on fusion of CSA and HS

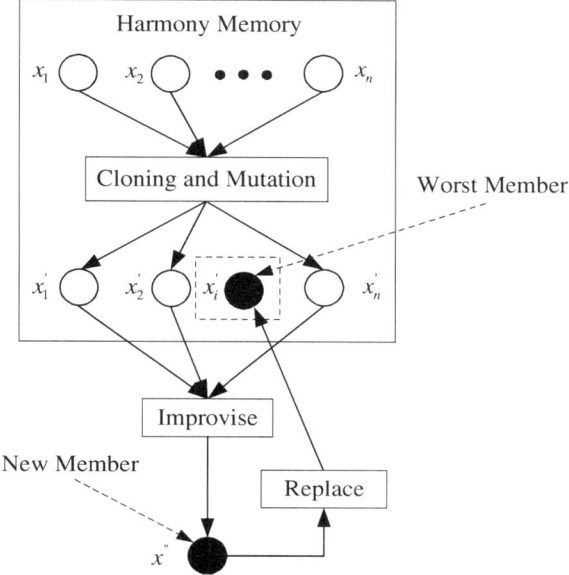

In the next section, we will demonstrate the enhanced performance of this hybrid
HS algorithm over the CSA and HS method in the optimization of fuzzy classification systems.

5.3.2 Hybrid HS Method in Optimization of Fuzzy Classification Systems

5.3.2.1 Fuzzy Classification Systems

Generally, in an n-input single-output fuzzy classification system, a representative
classification rule is as follows:

Rule l : IF x_1 is A_1^l and x_2 is A_2^l and ... and x_n is A_n^l, THEN y is C_m,

where $l = 1, \ldots, L$, L is the number of fuzzy rules, $m = 1, \ldots, M$, M is the number
of data classes, n is the number of input variables, and $A_j^l (j = 1, 2, \ldots, n)$ is a fuzzy
set associated with feature variable x_j. Here, vector $X = \begin{bmatrix} x_1, x_2, \ldots, x_n \end{bmatrix}$ in the
antecedent part consists of the input variables, and C_m in the consequent part is the
class label. We only consider the asymmetric triangular membership function for
those input variables:

$$\mu(x; a, b, c) = \max\left(0, \min\left(\frac{x-a}{b-a}, \frac{c-x}{c-b}\right)\right), \qquad (5.38)$$

where a, b, and c are the adaptive membership function parameters. Based on the given input data, the initial fuzzy system with a set of predefined rules according to the number of data classes can be obtained using some data clustering algorithms. The fuzzy c-means clustering method is a popular data clustering technique, which groups data or objects with high similarity and generates the partitions so that each object belongs to one or more clusters. In other words, it allows a data object to be classified into several clusters with different membership degrees. Furthermore, the parameters of the membership functions associated with the fuzzy sets can be optimized by any optimization methods.

In our fuzzy data classification scheme, we select the Sugeno fuzzy system with the singleton consequents representing different data classes. To evaluate the performance of the optimized membership functions, an objective function is defined as follows:

$$J = \frac{1}{K} \sum_{k=1}^{K} e_k, \tag{5.39}$$

where K is the number of the data samples in the training set and e_k is the classification error of a given data pattern. e_k is calculated as follows:

$$e_k = \begin{cases} 0, & \text{if classification is correct} \\ 1, & \text{if classification is incorrect.} \end{cases} \tag{5.40}$$

Therefore, the task of the proposed hybrid HS optimization method is to optimize the membership functions in the above Sugeno fuzzy classification system by minimizing the objective function so that its data classification rate can be maximized. In the next section, the Fisher iris data and wine data are used as two representative test beds for examining this approach.

5.3.2.2 Simulations

Fisher Iris Data Classification

The Fisher iris data are a well-known challenging benchmark for the data classification techniques, which consist of four input measurements, sepal length (SL), sepal width (SW), petal length (PL), and petal width (PW), in 150 data sets [17]. A total of three species are involved, which are Setosa, Versicolor, and Virginica, and each species contains 50 samples. To perform the iris data classification based on the output y of our Sugeno fuzzy classification system, the following principles are deployed:

$$\text{Iris} = \begin{cases} \text{Sestosa,} & \text{if } y < 0.4 \\ \text{Versiolor,} & \text{if } 0.4 \leq y \leq 0.9 \\ \text{Virginica,} & \text{if } y > 0.9. \end{cases} \tag{5.41}$$

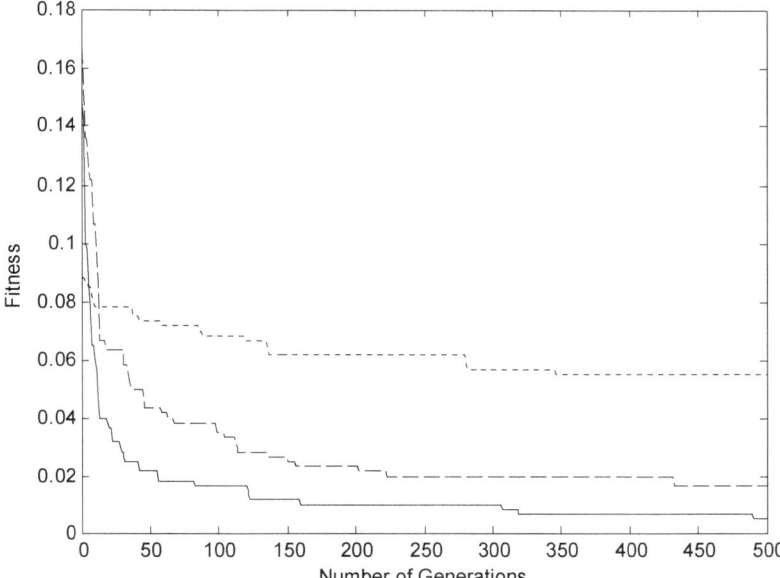

Fig. 5.12 Convergence procedures of CSA, HS, and hybrid HS algorithm in optimization of Sugeno fuzzy classification system for Fisher iris data classification ($N = 10$). *Dotted line* CSA, *dash line* HS method, *solid line* hybrid HS algorithm

In the simulations, N instances from each iris species are randomly selected as the training data, and the remaining instances (T) are regarded as the test data. All the input variables are normalized within the range of [0, 1]. For the hybrid HS optimization algorithm, we set HCMR = 0.8, PAR = 0.8, the number of the HM members is five, and the maximum number of the antibody clones is four. The CSA, HS, and proposed HS optimization algorithm are applied to optimize the aforementioned Sugeno fuzzy classification system. Figure 5.12 illustrates the performance comparison of their convergence speeds. Here, $N = 10$, and the test data have a total of 120 individual sets. Note that the results in Fig. 5.12 are the average of 10 independent runs. Obviously, our hybrid HS method can achieve the smallest classification error, and the corresponding classification rate is 99.2 %.

In a typical trial, a Sugeno fuzzy classification system with three rules is optimized, which results in only two misclassifications. The following three rules are available:

1. IF SL is Small and SW is Large and PL is Small and PW is Small, THEN iris is Setosa.
2. IF SL is Large and SW is Small and PL is Medium and PW is Medium, THEN iris is Versicolor.
3. IF SL is Medium and SW is Small and PL is Large and PW is Large, THEN iris is Virginica.

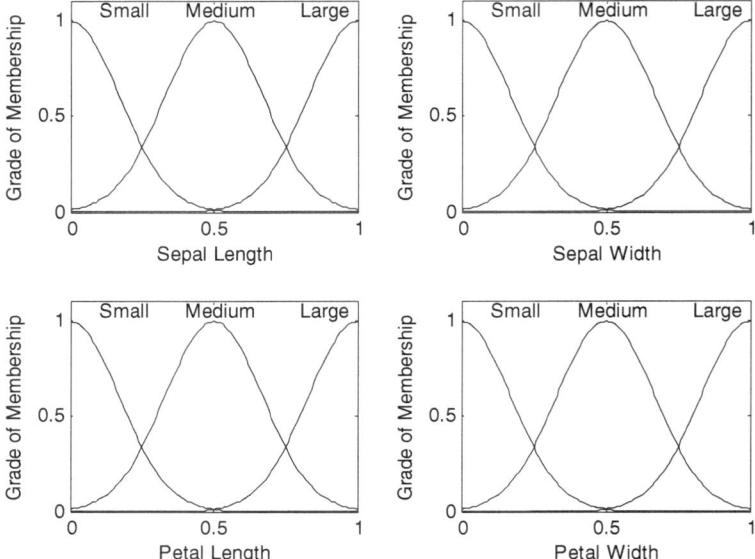

Fig. 5.13 Initial membership functions of Sugeno fuzzy classification system

Figures 5.13 and 5.14 show the initial and optimized membership functions of Small, Medium, and Large, respectively.

We further simplify our Sugeno fuzzy classification system by using only three input features, i.e., SW, PL, and PW, and assigning two membership functions to each of them. The three new fuzzy classification rules are as follows:

1. IF SW is Large and PL is Small and PW is Small, THEN iris is Setosa.
2. IF SW is Small and PL is Small and PW is Small, THEN iris is Versicolor.
3. IF SW is Large and PL is Large and PW is Large, THEN iris is Virginica.

The optimal membership functions acquired by the hybrid HS method are shown in Fig. 5.15, and a classification rate of 99 % has been achieved in this case.

Moreover, we examine the effectiveness of our hybrid HS algorithm in the optimization of the same Sugeno fuzzy classification system with different numbers of training data sets, as given in Table 5.6. The results are also the average of 10 separate runs. Misclassifying the patterns of Virginica into Versicolor is the main factor affecting the overall recognition rate, and the classification of Setosa is nearly 100 % correct. Additionally, the fuzzy rules extracted by the fuzzy c-means clustering method can influence the classification rate. Table 5.7 gives the performance comparison between our scheme and other existing solutions from several references. It is clearly visible that the Fisher iris data classification rate of the Sugeno fuzzy classification system can be significantly improved with the hybrid HS method.

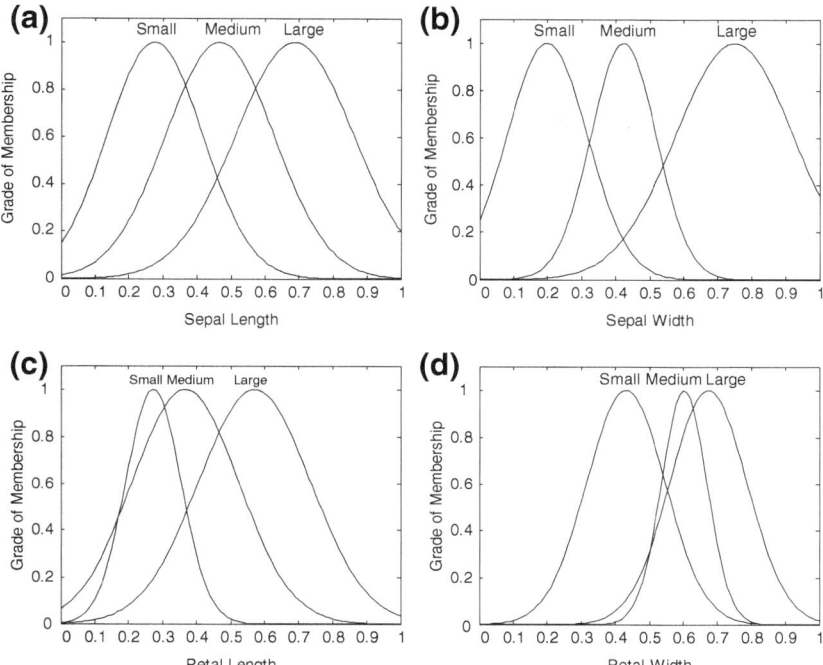

Fig. 5.14 Membership functions of Sugeno fuzzy classification system optimized by hybrid HS algorithm for Fisher iris data classification. **a** Membership functions of SL. **b** Membership functions of SW. **c** Membership functions of PL. **d** Membership functions of PW

Wine Data Classification

The wine data contain the chemical analysis of 178 wines that are brewed in the same region of Italy, but derived from three different cultivars. Each pattern consists of 13 features: alcohol content (Alc), malic acid content (Mal), ash content, alkalinity of ash (Ash), magnesium content (Mag), total phenols (Tot), flavanoids (Fla), non-flavanoids phenols (nFlav), proanthocyaninsm (Proa), color intensity (Col), hue, OD280/OD315 (OD2) of diluted wines, and praline (Pro). The numbers of the patterns in these three classes are 59, 71, and 48, respectively [18].

Like in the Fisher iris data classification, the output y of the Sugeno fuzzy classification system is based on the following classification rules:

$$\text{wine} = \begin{cases} \text{Class 1,} & \text{if } y < 0.33 \\ \text{Class 2,} & \text{if } 0.33 \leq y \leq 0.67 \\ \text{Class 3,} & \text{if } y > 0.67. \end{cases} \tag{5.42}$$

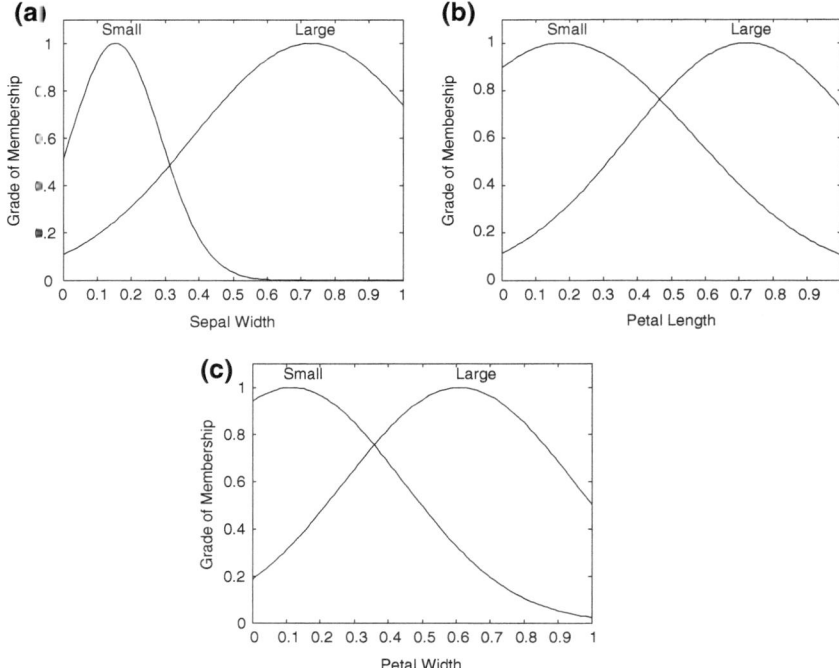

Fig. 5.15 Membership functions of simplified Sugeno fuzzy classification system optimized by hybrid optimization algorithm for Fisher iris data classification. **a** Membership functions of SW. **b** Membership functions of PL. **c** Membership functions of PW

Table 5.6 Classification results of Fisher iris data using CSA, HS, and hybrid method

Algorithms	CSA	HS	Hybrid HS method
$N = 10, T = 40$			
Misclassifications (training)	3	2	0.2
Classification rate (training)	90 %	93.3 %	99.3 %
Misclassifications (test)	6.4	4.2	1
Classification rate (test)	94.7 %	96.5 %	99.2 %
$N = 20, T = 30$			
Misclassifications (training)	3.8	3.2	0.2
Classification rates (training)	93.7 %	94.7 %	99.5 %
Misclassifications (test)	3.4	3	0.6
Classification rates (test)	96.2 %	96.7 %	99.3 %
$N = 40, T = 10$			
Misclassifications (training)	4.2	3.8	0.6
Classification rates (training)	96.5 %	96.8 %	99.5 %
Misclassifications (test)	1.6	1.4	0.2
Classification rates (test)	94.7 %	95.3 %	99.3 %

Table 5.7 Fisher iris data classification comparisons of results from different references

References	Number of features	Number of rules	Classification rates (%)
Shi et al. [30]	12	4	98
Setnes and Roubos [29]	8 and 12	2 and 3	99.3 and 98.9
Russo [31]	18	5	100
Chang and Lilly [18]	7	5	99.3
Hybrid HS method	6 and 12	3	99 and 99.3

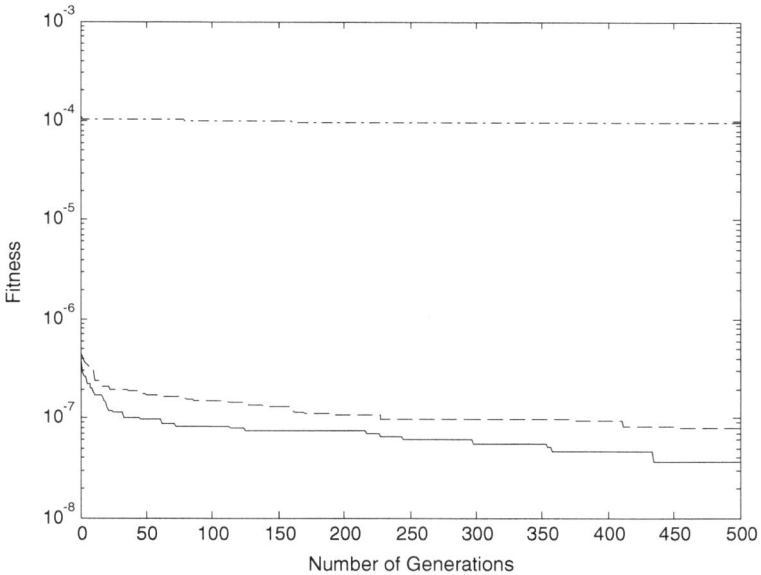

Fig. 5.16 Convergence procedures of CSA, HS, and hybrid HS algorithm in optimization of fuzzy classification system for wine data classification ($N = 20$). *Dash-dotted line* CSA, *dash line* HS, *solid line* hybrid HS algorithm

Figure 5.16 illustrates the convergence speed comparison among the CSA, HS, and proposed hybrid HS algorithm. The results are the average of 10 independent runs. $N = 20$, and 118 sets of wine data are used as the test data. As we can observe, the hybrid HS method yields the best classification performance.

As an illustrative example, the initial membership functions of flavanoids and color intensity in the simplified Sugeno fuzzy classification system are demonstrated in Fig. 5.17. Figure 5.18 shows the optimized membership functions of Small, Medium, and Large. The following three rules are utilized:

1. IF Mal is Small and Tot is Large and Fla is Large and Col is Medium and Hue is Large and OD2 is Large and Pro is Large, THEN wine is Class 1.

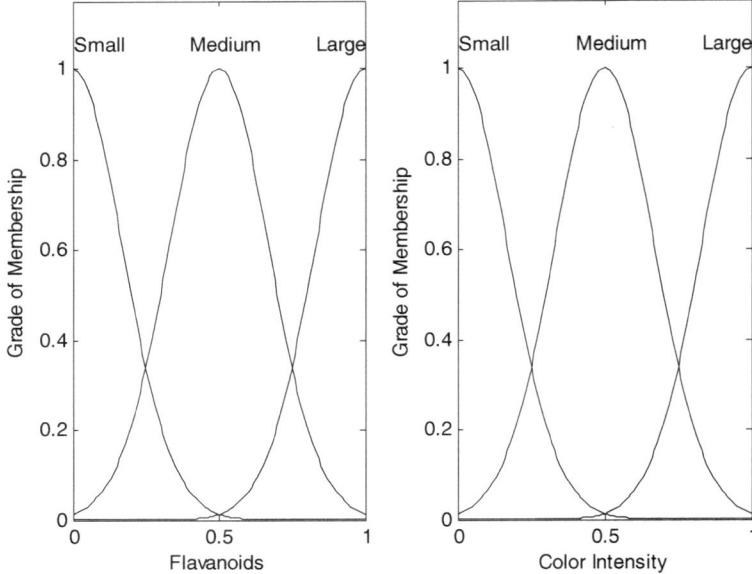

Fig. 5.17 Initial membership functions of flavanoids and color intensity in fuzzy classification system for wine data classification

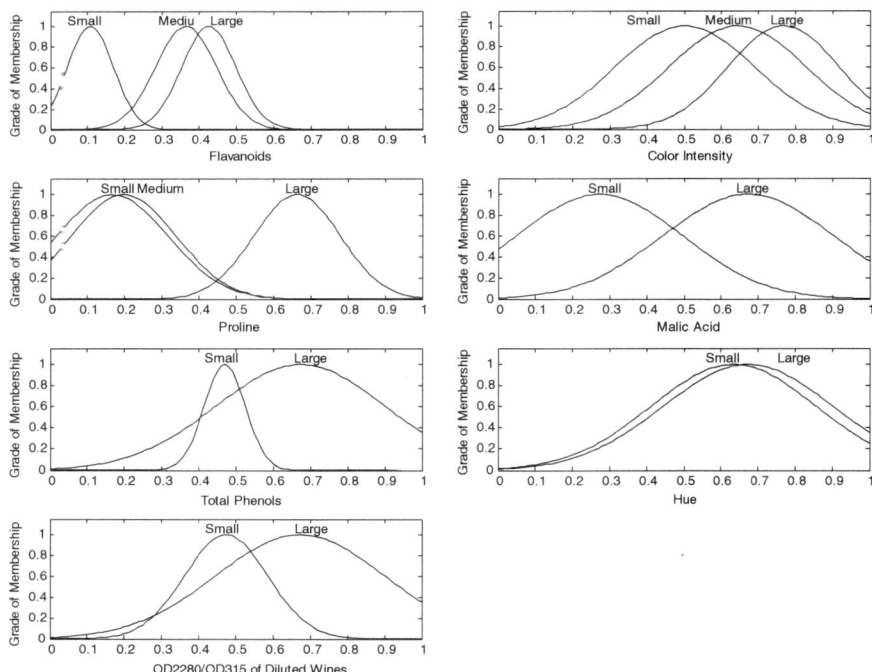

Fig. 5.18 Membership functions of simplified Sugeno fuzzy classification system optimized by hybrid HS algorithm for wine data classification

Table 5.8 Classification results of wine data using CSA, HS, and hybrid HS method

Algorithms	CSA	HS	Hybrid HS method
$N = 10$			
Misclassifications (training)	12	10	3
Classification rate (test)	93.2 %	94.3 %	98.3 %
$N = 30$			
Misclassifications (training)	10	9	1
Classification rates (test)	94.3 %	94.9 %	99.4 %

2. IF Mal is Small and Tot is Small and Fla is Medium and Col is Small and Hue is Large and OD2 is Large and Pro is Small, THEN wine is Class 2.
3. IF Mal is Large and Tot is Small and Fla is Small and Col is Large and Hue is Small and OD2 is Small and Pro is Medium, THEN wine is Class 3.

Similarly, we explore the efficiency of our hybrid HS algorithm in the optimization of the same Sugeno fuzzy classification system with different numbers of training data sets, as given in Table 5.8. The results here are the average of 10 separate trials as well. Compared with both the CSA and HS, employment of the proposed hybrid HS optimization method leads to the optimal wine data classification results.

5.4 A Hybrid HS Method with Cultural Algorithm

It has been proved that the employment of an effective mechanism to represent, acquire, store, and utilize the search knowledge can bias and accelerate the convergence of various evolutionary computation methods [19]. As a matter of fact, Reynolds proposes a distinguishing cultural algorithm (CA), in which there exist two spaces, population space and belief space, for the acquisition and deployment of search knowledge [20]. The problem-solving experiences from the individuals in the population space are first extracted and stored in the belief space and are next used to influence the evolution of the population space. Therefore, in this section, we develop and explore a hybridization of the HS and CA: HS-CA. In our HS-CA, the situational and normative knowledge from the CA is applied to properly adjust and guide the mutation of the new solution candidates in the HS method so that an improved convergence performance can be achieved. The application of the proposed HS-CA in the optimal design of wind generator is also investigated [6].

5.4.1 Cultural Algorithm

As proposed by Reynolds, the CA is a dual inheritance system with its evolution on two different levels: population level and belief level [21]. The culture in the CA refers to the beliefs or experiences, which can be gained from and used to direct the evolving individuals. The basic framework of the CA is shown in Fig. 5.19, where there are two spaces, population space and belief space, interacting with each other. Similar to the regular evolutionary computation algorithms, the evolution of the individuals in the population space is based on such operations as mutation and crossover [22, 23]. However, besides the population space, the CA has the belief space that can acquire and store certain domain knowledge from the individuals in the population space. More precisely, the most fitted individuals in the population space are selected by the 'Accept' function to update the knowledge in the belief space. The function of 'Influence' takes advantage of this kind of knowledge to guide the evolution of the population space. The knowledge in the belief space is classified into two essential types: situational knowledge and normative knowledge. The situational knowledge is actually the best exemplar chosen from the population space, while the normative knowledge contains the useful information concerning the search regions, where the above-average individuals may exist. In a word, both the situational knowledge and normative knowledge in the belief space can control and improve the population space.

The pseudo-codes of the aforementioned CA can be presented as follows [19]:

Step 1:   Set $t = 0$.
Step 2:   Initialize population space P^t.
Step 3:   Initialize belief space B^t.

Fig. 5.19 Cultural algorithm (*CA*)

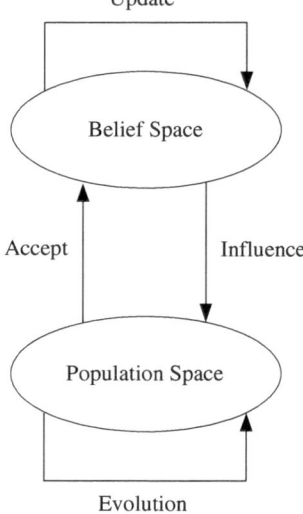

Step 4: Update belief space B^t using Accept(P^t).
Step 5: Update population space P^t using Influence(B^t).
Step 6: $t = t + 1$.
Step 7: Return back to Step 4 until a given termination criterion is met.

It is concluded from the above descriptions that the unique belief space plays a key role in the CA. In the next section, we will integrate the knowledge management and utilization mechanisms of the CA into the original HS method and propose a hybrid optimization approach: HS-CA. This HS-CA uses the belief space to enhance the generation of new HM members so that their fitness can be significantly improved.

5.4.2 Fusion of HS and CA: HS-CA

We can observe that based on the PAR, the new solution candidates improvised from the existing HM members may go through a mutation procedure. The mutation is generally a 'blind' and local exploration in the search space. Unfortunately, it is well known that the pure random-based mutation is not effective in guiding the evolution of individuals. Thus, the CA can be employed to influence the direction and size of the mutation in the HS method. That is to say, the knowledge in the belief space of the CA is first extracted from the HM and next used to direct the mutation of the new solutions. Inspired by this idea, we here propose a hybridization of the HS method and CA: HS-CA, as illustrated in Fig. 5.20.

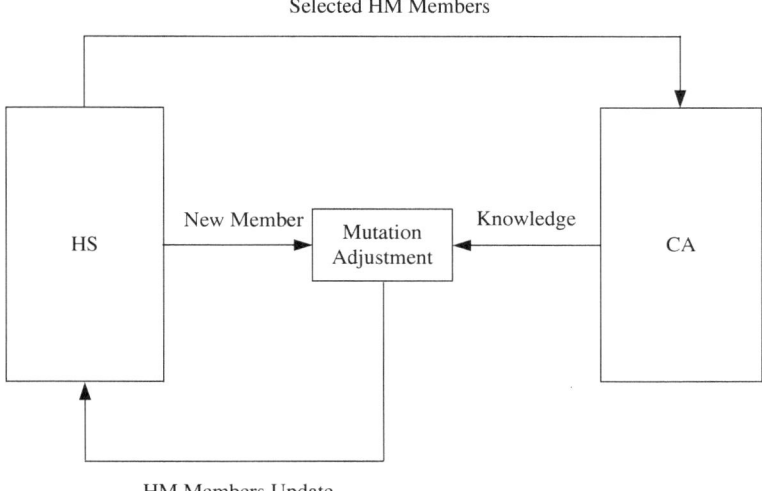

Fig. 5.20 Hybridization of HS and CA: HS-CA

To put it into more details, suppose any individual in the HS-CA is represented as $\mathbf{x}_i^t = \left[x_{i,1}^t, x_{i,2}^t, \ldots, x_{i,N}^t \right]$, where N is the dimension of the solution space and t is the iteration step. All the individuals are evaluated by fitness function $f(\cdot)$. The optimization goal is, therefore, to find the optimal $\mathbf{x}^t$ that can maximize $f(\cdot)$. The situational knowledge of the CA is denoted as $\mathbf{S}^t$. As discussed above, $\mathbf{S}^t$ should be the best individual selected from the HM and is updated as follows:

$$\mathbf{S}^{t+1} = \mathbf{x}_{\text{best}}^t, \quad \text{if } f\left(\mathbf{x}_{\text{best}}^t\right) > f(\mathbf{S}^t),$$
$$\mathbf{S}^{t+1} = \mathbf{S}^t, \qquad \text{otherwise.}$$

The normative knowledge of the CA is represented by N intervals for the dimensions of the individuals. Each interval is associated with a triple $\langle I_i^t, L_i^t, U_i^t \rangle$. I_i^t is a closed interval with lower bound l_i^t and upper bound u_i^t. L_i^t and U_i^t are the 'scores' for l_i^t and u_i^t, respectively. Usually, l_i^t and u_i^t are initialized by the given problem domain, and both L_i^t and U_i^t are initially set to be $+\infty$. $\langle I_i^t, L_i^t, U_i^t \rangle$ can be updated by the top HM members $\mathbf{x}_1^t, \mathbf{x}_2^t, \ldots, \mathbf{x}_M^t$, where M is the number of the selected candidates, e.g., $M = \text{HMS} \times \text{Top}$, where Top is a percentage constant. For $\mathbf{x}_j^t$, we have

$$l_i^{t+1} = x_{j,i}^t, \qquad \text{if } x_{j,i}^t \leq l_i^t \quad \text{or } f\left(\mathbf{x}_j^t\right) > L_i^t,$$
$$l_i^{t+1} = l_i^t, \qquad \text{otherwise.}$$
$$L_i^{t+1} = f\left(\mathbf{x}_j^t\right), \qquad \text{if } x_{j,i}^t \leq l_i^t \quad \text{or } f\left(\mathbf{x}_j^t\right) > L_i^t,$$
$$L_i^{t+1} = L_i^t, \qquad \text{otherwise.}$$
$$u_i^{t+1} = x_{j,i}^t, \qquad \text{if } x_{j,i}^t \geq u_i^t \quad \text{or } f\left(\mathbf{x}_j^t\right) > U_i^t,$$
$$u_i^{t+1} = u_i^t, \qquad \text{otherwise.}$$
$$U_i^{t+1} = f\left(\mathbf{x}_j^t\right), \qquad \text{if } x_{j,i}^t \geq u_i^t \quad \text{or } f\left(\mathbf{x}_j^t\right) > U_i^t,$$
$$U_i^{t+1} = U_i^t, \qquad \text{otherwise.}$$

Note that the update policy of the normative knowledge is that it should be conservative and progressive to narrow and widen, respectively, the interval I_i^t. Both the situational knowledge and normative knowledge are utilized to determine the mutation operation in the HS of our HS-CA.

Suppose $\mathbf{x}_*$ is a new solution improvised by the random combination of the components of the HM members and $\mathbf{x}'_*$ is the mutated version of $\mathbf{x}_*$. If a conventional mutation operator is used, $\mathbf{x}'_*$ can be generated as follows:

$$\mathbf{x}'_* = \mathbf{x}_* + \text{Rand}, \tag{5.43}$$

where Rand represents a random number generator. However, compared with (2), the Influence function of the CA on the mutation is more goal-directed. There are four basic types of Influence functions, N_s, S_d, $N_s + S_d$, and $N_s + N_d$, depending on

the ways how the situational knowledge and normative knowledge are employed
to guide the mutation. They are explained in details as follows.
For N_s,

$$\mathbf{x}'_{*i} = \mathbf{x}_{*i} + \text{Size}(I_i) \times N(0, 1), \tag{5.44}$$

where $\text{Size}(I_i)$ is the size of the belief interval I_i for the ith dimension of $\mathbf{x}_*$ and
$N(0, 1)$ is a random number with the normal distribution, whose mean and stan-
dard derivation are 0 and 1, respectively.
For S_d,

$$
\begin{aligned}
\mathbf{x}'_{*i} &= \mathbf{x}_{*i} + |\sigma_i \times N(0, 1)|, && \text{if } \mathbf{x}_{*i} < \mathbf{S}_i, \\
\mathbf{x}'_{*i} &= \mathbf{x}_{*i} - |\sigma_i \times N(0, 1)|, && \text{if } \mathbf{x}_{*i} > \mathbf{S}_i, \\
\mathbf{x}'_{*i} &= \mathbf{x}_{*i} + \sigma_i \times N(0, 1), && \text{otherwise,}
\end{aligned}
$$

where σ_i is the given mutation step for the ith dimension of $\mathbf{x}_*$ and $\mathbf{S}_i$ is the
situational knowledge for the ith dimension of $\mathbf{x}_*$ in the belief space.
For $N_s + S_d$,

$$
\begin{aligned}
\mathbf{x}'_{*i} &= \mathbf{x}_{*i} + |\text{Size}(I_i) \times N(0, 1)|, && \text{if } \mathbf{x}_{*i} < \mathbf{S}_i, \\
\mathbf{x}'_{*i} &= \mathbf{x}_{*i} - |\text{Size}(I_i) \times N(0, 1)|, && \text{if } \mathbf{x}_{*i} > \mathbf{S}_i, \\
\mathbf{x}'_{*i} &= \mathbf{x}_{*i} + \text{Size}(I_i) \times N(0, 1), && \text{otherwise.}
\end{aligned}
$$

For $N_s + N_d$,

$$
\begin{aligned}
\mathbf{x}'_{*i} &= \mathbf{x}_{*i} + |\text{Size}(I_i) \times N(0, 1)|, && \text{if } \mathbf{x}_{*i} < \mathbf{S}_i, \\
\mathbf{x}'_{*i} &= \mathbf{x}_{*i} - |\text{Size}(I_i) \times N(0, 1)|, && \text{if } \mathbf{x}_{*i} > \mathbf{S}_i, \\
\mathbf{x}'_{*i} &= \mathbf{x}_{*i} + \beta \times \text{Size}(I_i) \times N(0, 1), && \text{otherwise,}
\end{aligned}
$$

where l_i and u_i are the normative knowledge (lower and upper bounds for the ith
dimension of $\mathbf{x}_*$, respectively) in the belief space and β is a preset small constant.

It should be emphasized that the belief space in the CA can identify the
promising regions, where the above-average individuals may be present. The
objective of the Influence function is to promote the newly generated solutions to
those regions so that their fitness is improved. Different from the mutation operation in (5.32), the one of the proposed HS-CA is effectively controlled by the
Influence function. Therefore, embedding the CA into the HS method can indeed
yield a superior optimization performance. In the following section, we are going
to use 15 nonlinear functions and four engineering problems to demonstrate the
accelerated convergence of our HS-CA.

5.4.3 Optimization of Nonlinear Functions and Engineering Design Using HS-CA

In this section, we investigate the effectiveness of the proposed HS-CA with some simulation examples of nonlinear functions as well as engineering problems.

5.4.3.1 Nonlinear Functions

Fifteen n-dimensional nonlinear functions (12 mentioned in Sect. 5.2.3), which have been widely used as popular optimization benchmarks [8, 24], are employed to compare the optimization (minimization) capabilities between the HS and HS-CA. In addition to the functions in Sect. 5.2.3, the following three functions are considered as well.

Dixon and Price function:

$$f(\mathbf{x}) = (x_1 + 1)^2 + \sum_{i=1}^{n} i\left(2x_i^2 - x_{i-1}\right)^2, \quad x_i \in [-10, 10]. \tag{5.45}$$

Michalewicz function:

$$f(\mathbf{x}) = -\sum_{i=1}^{n} \sin(x_i)\left[\sin\left(\frac{ix_i^2}{\pi}\right)\right]^{20}, \quad x_i \in [0, \pi]. \tag{5.46}$$

Zakharov function:

$$f(\mathbf{x}) = \sum_{i=1}^{n} x_i^2 + \left(\sum_{i=1}^{n} \frac{ix_i}{2}\right)^2 + \left(\sum_{i=1}^{n} \frac{ix_i}{2}\right)^4, \quad x_i \in [-5, 10]. \tag{5.47}$$

The global minima of all these functions are at $f(\mathbf{x}) = 0$, except for Michalewicz function and Trid function, whose global minima are unknown when $n = 50$.

Again, we use the NFE as the principal criterion to compare the convergence speeds of the HS and HS-CA. Both of them have 100 HM members, i.e., HMS $= 100$, which are always initialized to be equal. The relevant parameters in these two optimization methods are as follows: HMCR $= 0.8$ and PAR $= 0.6$. Their evolution procedures are terminated after 10,000 NFE. In the HS-CA, we use the Influence function of $N_s + N_d$ for all the 15 functions, and the corresponding CA parameters of Top and β are given in Table 5.9.

Tables 5.10, 5.11 and 5.12 present the optimal solutions acquired by the HS and HS-CA, when $n = 10$, $n = 30$, and $n = 50$, respectively. We stress that the results here are based on the average of 100 independent trials. As an illustrative example, the optimal solutions to the Ackley function ($n = 10$) acquired from the HS and

Table 5.9 Parameters of Top and β in HS-CA

	Top (%)	β
Ackley function	10	0.1
Alpine function	10	0.1
Bohachevsky function	10	0.1
De Jong function	10	0.1
Dixon and Price function	10	0.1
Griewank function	10	0.1
Hyper-ellipsoid function	10	0.1
Levy function	10	0.1
Michalewicz function	10	0.01
Powell function	10	0.1
Rastrigin function	5	0.1
Rosenbrock function	10	0.1
Sphere function	10	0.1
Trid function	10	0.2
Zakharov function	5	0.1

Table 5.10 Optimal solutions acquired by HS and HS-CA within 10,000 NFE ($n = 10$)

	HS	HS-CA
Ackley function	1.4802	1.3516×10^{-12}
Alpine function	0.0117	2.1185×10^{-4}
Bohachevsky function	0.7144	0.0105
De Jong function	8.8887×10^{-6}	2.1577×10^{-45}
Dixon and Price function	1.4288	0.6613
Griewank function	21.6018	0.1543
Hyper-ellipsoid function	137.1776	0.0962
Levy function	0.0944	6.5504×10^{-24}
Michalewicz function	-9.0838	-9.5722
Powell function	0.0378	0.0015
Rastrigin function	2.8585	0.2498
Rosenbrock function	44.4376	7.8584
Sphere function	0.0025	2.7335×10^{-24}
Trid function	-165.7292	-200.5229
Zakharov function	2.5722	8.5050×10^{-4}

HS-CA are shown in Fig. 5.21a, b, respectively. In addition, Fig. 5.22 shows the iteration procedures (average over 100 trials) of these two methods in optimizing the same function. Apparently, compared with the original HS method, for all the

Table 5.11 Optimal solutions acquired by HS and HS-CA within 10,000 NFE ($n = 30$)

	HS	HS-CA
Ackley function	2.5364	5.165×10^{-5}
Alpine function	0.5881	0.0015
Bohachevsky function	4.6559	0.3680
De Jong function	0.0046	8.2869×10^{-17}
Dixon and Price function	24.5650	0.8037
Griewank function	61.9223	0.3622
Hyper-ellipsoid function	5.6404×10^8	1.1386×10^7
Levy function	1.1288	0.0398
Michalewicz function	-17.2730	-28.6997
Powell function	3.7738	0.1544
Rastrigin function	24.0671	11.3442
Rosenbrock function	251.3008	35.4433
Sphere function	0.0773	6.0406×10^{-10}
Trid function	-451.8707	-897.9181
Zakharov function	261.7912	176.3173

Table 5.12 Optimal solutions acquired by HS and HS-CA within 10,000 NFE ($n = 50$)

	HS	HS-CA
Ackley function	3.0282	0.0126
Alpine function	2.3841	0.0078
Bohachevsky function	12.4675	1.4351
De Jong function	10.0609	6.7928×10^{-9}
Dixon and Price function	415.4205	3.3149
Griewank function	102.9463	3.2802
Hyper-ellipsoid function	5.7603×10^{14}	3.2679×10^{11}
Levy function	3.2569	0.1956
Michalewicz function	-21.9800	-44.0748
Powell function	14.2161	1.1277
Rastrigin function	61.6260	39.7626
Rosenbrock function	821.9302	91.6834
Sphere function	0.6496	1.4131×10^{-5}
Trid function	-185.4805	-996.127
Zakharov function	581.7118	470.9657

15 test functions, our HS-CA can achieve considerably better optimization results within the same NFE, due to the CA-based utilization of the search knowledge extracted from the HS. That is, the HS-CA has a superior nonlinear function optimization capability.

Fig. 5.21 Optimal solutions
to Ackley function ($n = 10$)
acquired by HS and HS-CA.
a HS and **b** HS-CA

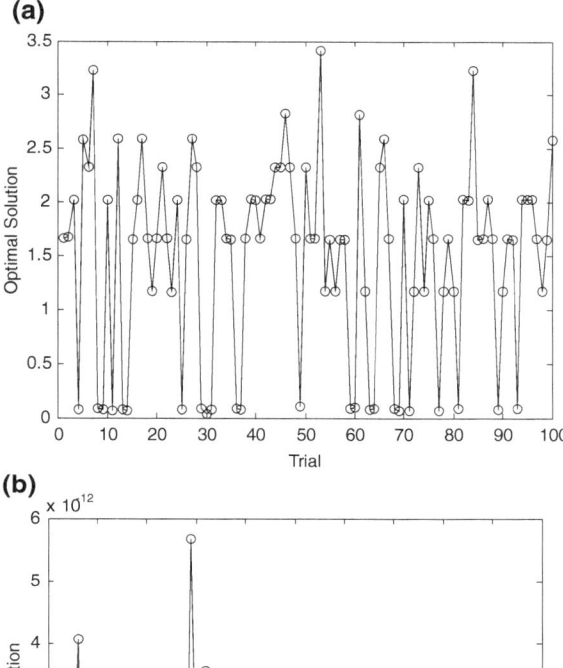

Fig. 5.22 Optimization
procedures of Ackley
function ($n = 10$) with HS
(*solid line*) and HS-CA
(*dotted line*)

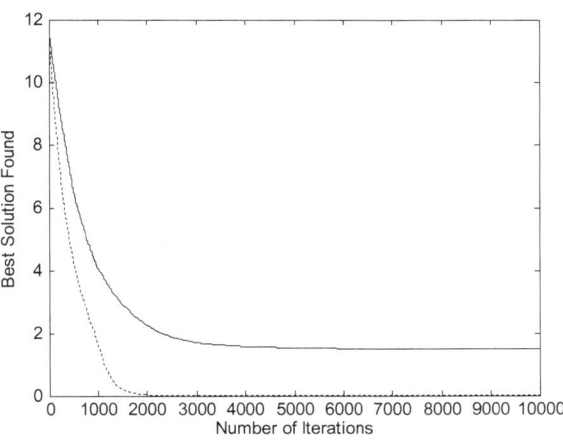

Fig. 5.23 Optimization procedures of Rastrigin function ($n = 50$) with HS-CA (*solid line* Top = 5 %, *dotted line* Top = 10 %, *dashed line* Top = 15 %)

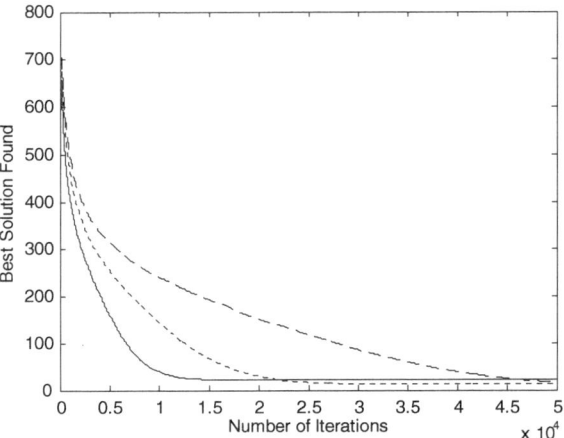

The above simulation results indeed demonstrate that the HS-CA can outperform the regular HS method. Nevertheless, its performance is unavoidably affected by the CA parameters involved. More specifically, the choice of Top, β, and Influence function play a pivotal role in the optimization capability of the HS-CA. For example, Fig. 5.23 illustrates the iteration procedures of the HS-CA with Top = 5 %, Top = 10 %, and Top = 15 % for the optimization of the Rastrigin function ($n = 50$). Apparently, different Top values can lead to different convergence characteristics of our HS-CA. We also examine the effect of parameter β on the HS-CA in optimizing the Michalewicz function ($n = 50$), as shown in Fig. 5.24. The HS-CA with $\beta = 0.01$ converges must faster than the ones with $\beta = 0.05$ and $\beta = 0.1$. Unfortunately, there is no analytic way yet for us to choose the best values for these CA parameters, which are often determined based on *trial and error*. The four kinds of the Influence functions for the Rastrigin function

Fig. 5.24 Optimization procedures of Michalewicz function ($n = 50$) with HS-CA (*solid line* $\beta = 0.1$, *dotted line* $\beta = 0.05$, *dashed line* $\beta = 0.01$)

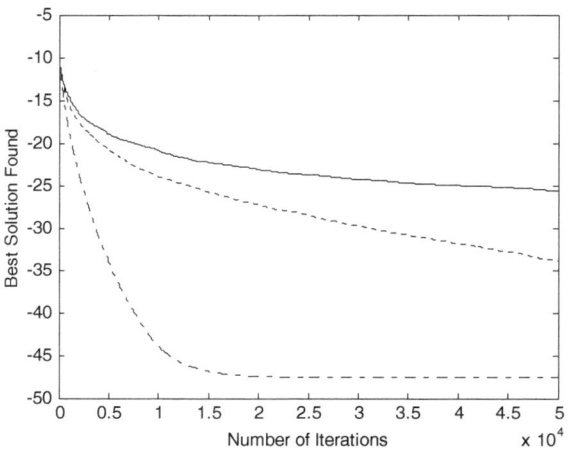

Fig. 5.25 Optimization
procedures of Rastrigin
function ($n = 50$) with
HS-CA (*solid line* N_s, *dotted
line* S_d, *dashed-dotted line*
$N_s + S_d$, *dashed line*
$N_s + N_d$)

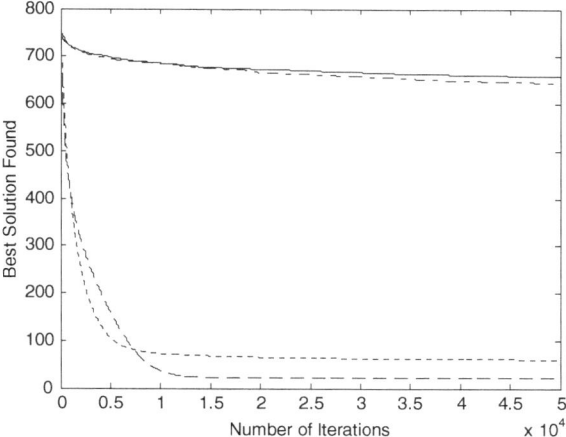

optimization are explored in Fig. 5.25. It can be observed that the effectiveness of
this HS-CA significantly deteriorates with the Influence functions of N_s and $N_s +
S_d$. Reynolds argues that the reason behind this may lie in the structures of the
functions to be optimized [20]. Further discussions on the Influence function are
given in the same paper as well.

5.4.3.2 Engineering Optimization Problems

Example 1 Optimal welded beam design

The optimal design of the welded beam has been an important benchmark for
modern optimization methods [25]. The goal here is to minimize the fabricating
cost of the welded beam, subject to the constraints on the shear stress of the weld,
$\tau(\mathbf{x})$, bending stress on the beam, $\sigma(\mathbf{x})$, buckling load on the beam, $P_C(\mathbf{x})$, end
deflection of the beam, $\delta(\mathbf{x})$, and side constraints. The four design variables, h, l, t,
and b, are denoted as x_1, x_2, x_3, and x_4, respectively, as illustrated in Fig. 5.26. The
details of this practical optimal welded beam design problem are as follows:

$$f(\mathbf{x}) = (1 + c_1)x_1^2 x_2 + c_2 x_3 x_4 (x_2 + 14),$$

subject to $\quad g_1(\mathbf{x}) = \tau(\mathbf{x}) - \tau_{\max} \leq 0,$

$$g_2(\mathbf{x}) = \sigma(\mathbf{x}) - \sigma_{\max} \leq 0,$$

$$g_3(\mathbf{x}) = x_1 - x_4 \leq 0,$$

$$g_4(\mathbf{x}) = \delta(\mathbf{x}) - \delta_{\max} \leq 0,$$

$$g_5(\mathbf{x}) = F - P_C(\mathbf{x}) \leq 0,$$

Fig. 5.26 Welded beam
design

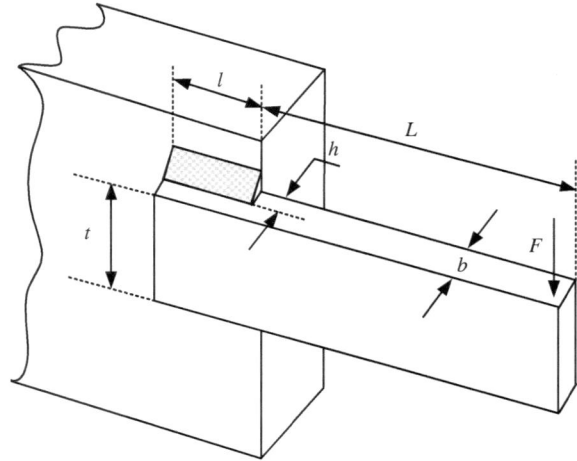

where c_1 $(c_1 = 0.10471)$ and c_2 $(c_2 = 0.04811)$ are the unit costs of the weld
material and bar stock, respectively, $\tau(\mathbf{x}) = \sqrt{(\tau')^2 + 2\tau'\tau'' \frac{x_2}{2R} + (\tau'')^2}$, $\tau' = \frac{F}{\sqrt{2}x_1 x_2}$,

$\tau'' = \frac{MR}{J}$, $M = F\left(L + \frac{x_2}{2}\right)$, $R = \sqrt{\frac{x_2^2}{4} + \left(\frac{x_1+x_3}{2}\right)^2}$, $J = 2\left\{\sqrt{2}x_1 x_2 \left[\frac{x_2^2}{12} + \left(\frac{x_1+x_3}{2}\right)^2\right]\right\}$,

$\sigma(\mathbf{x}) = \frac{6FL}{x_3^2 x_4}$, $\delta(\mathbf{x}) = \frac{4FL^3}{Ex_3^3 x_4}$, $P_C(\mathbf{x}) = \frac{4.013E\sqrt{\frac{x_3^2 x_4^6}{36}}}{L^2}\left(1 - \frac{x_3}{2L}\sqrt{\frac{E}{4G}}\right)$, $F = 6{,}000$ lb,

$L = 14$ in, $\delta_{max} = 0.25$ in, $E = 30 \times 10^6$ psi, $G = 12 \times 10^6$ psi, $\tau_{max} = 13{,}600$
psi, $\sigma_{max} = 30{,}000$ psi, $0.125 \leq x_1 \leq 5$, $0.1 \leq x_2, x_3 \leq 10$, and $0.1 \leq x_4 \leq 5$.

   Again, we use the same parameters as in the function optimization case, except
that NFE $= 100{,}000$, and β is chosen to be $\beta = 0.1$. Table 5.13 shows the average
optimization results of our HS-CA and HS method. We can observe that the
former is capable of outperforming the later in attacking this demanding welded
beam design problem.

Example 2 Optimal gear train design

   The goal of the optimal gear train design is to minimize the cost of the gear
ratio (GR) of a gear train, which is shown in Fig. 5.27 [8]. The GR can be defined
as follows:

$$GR = \frac{n_B n_D}{n_F n_A}, \tag{5.48}$$

where n_A, n_B, n_D, and n_F are the numbers of the teeth of the gearwheel in
Fig. 5.27, and they are denoted as x_1, x_2, x_3, and x_4, respectively, in this

Table 5.13 Optimization results of welded beam design with HS and HS-CA

	x_1	x_2	x_3	x_4	$f(\mathbf{x})$
HS	0.269560	4.114880	7.912975	0.305209	2.306814
HS-CA	0.299005	2.744191	7.502979	0.311244	2.093270

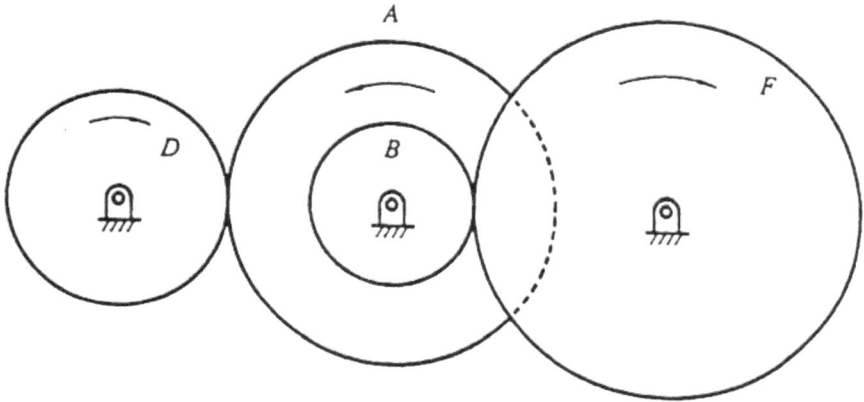

Fig. 5.27 Gear train design

Table 5.14 Optimization results of gear train design with HS and HS-CA

	x_1	x_2	x_3	x_4	$f(\mathbf{x})$
HS	47.91	20.12	17.63	48.66	9.3078×10^{-10}
HS-CA	48.94	19.14	18.73	49.31	2.4815×10^{-10}

engineering optimization problem. x_1, x_2, x_3, and x_4 are all integers between [12, 60]. The objective function $f(\mathbf{x})$ to be minimized is as follows:

$$f(\mathbf{x}) = \left(\frac{1}{6.931} - \frac{x_3 x_2}{x_1 x_4} \right). \tag{5.49}$$

The same simulation parameters as in the above example are used in the HS and HS-CA here. The *average* optimal design variables and costs acquired are given in Table 5.14, which actually demonstrate the superior optimization capability of our HS-CA.

Example 3 Optimal pressure vessel design

The structure of the center and end section of a pressure vessel is shown in Fig. 5.28, which is made of carbon steel ASME SA 203 grade B [26]. The objective of the optimal design is to find a feasible set of dimensions T_s (shell thickness), T_h (spherical head thickness), R (radius of cylindrical shell), and L

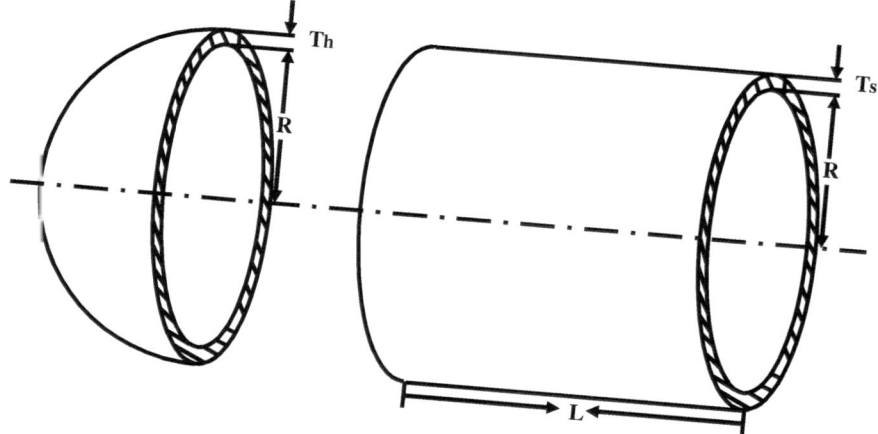

Fig. 5.28 Pressure vessel design

Table 5.15 Optimization results of pressure vessel design with HS and HS-CA

	x_1	x_2	x_3	x_4	$f(\mathbf{x})$
HS	1.8088	2.1281	57.0342	108.2850	2.3396×10^4
HS-CA	1.0781	1.0006	54.0931	69.1591	9.2040×10^3

(shell length) with a minimum total manufacturing cost for the pressure vessel, subject to the constraints on the minimal wall thicknesses, the minimal value of the tank, as well as the length of the cylindrical shell. The design variables T_s, T_h, R, and L are denoted as x_1, x_2, x_3, and x_4, respectively, here. Note that the variables T_s and T_h must be the integer multiples of 0.0625.

$$\begin{aligned}
\text{minimize} \quad & f(\mathbf{x}) = 0.6224x_1x_3x_4 + 1.7781x_2x_3^2 + 3.1611x_1^2x_4 + 19.84x_1^2x_3, \\
\text{subject to} \quad & g_1(\mathbf{x}) = 0.0193x_3 - x_1 \le 0, \\
& g_2(\mathbf{x}) = 0.00954x_3 - x_2 \le 0, \\
& g_3(\mathbf{x}) = 750 \times 1728 - \pi x_3^2 x_4 - 0.75\pi x_3^3 \le 0, \\
& g_4(\mathbf{x}) = x_4 - 240 \le 0, \\
& 1 \le x_1, x_2 \le 99, 10.0 \le x_3, x_4 \le 200.0.
\end{aligned}$$

The average-optimized pressure vessel design variables and costs obtained by the HS and HS-CA are provided in Table 5.15. Apparently, the optimization performance of the proposed HS-CA is moderately better than that of the original HS method in manipulating this optimal design problem.

Fig. 5.29 Cross section and
dimensions of permanent
magnet generator

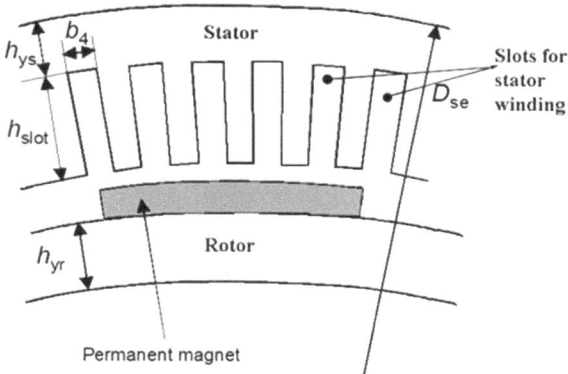

5.4.4 Optimal Design of Permanent Magnet Direct-Driven Wind Generator: A Case Study

5.4.4.1 Structure of Wind Generator

The wind generator shown in Figs. 5.29, 5.30 and 5.31 is a radial flux-type permanent magnet generator, in which the NdFeB magnets are surface-mounted. The remanence flux density of the magnets is 1.05 T and coercivity 800 kA/m. The stator winding is a three-phase two-layer full-pitch diamond winding. The number of the slots per pole and phase is 2. The stator slot form and the constant dimensions of the slot are illustrated in Fig. 5.30. The stator iron core consists of 55-mm-long subcores, between which there are radial 6-mm-wide ventilation ducts. The length of the subcore is constant, and the number of the ventilation ducts is a decimal fraction in the calculations. The stator frame, the bearing shields, and the rotor steel body are all 20 mm thick (Fig. 5.31). In the rotor body disk, there are holes, and around 50 % of the disk is iron and 50 % holes. The iron loss factor of the stator lamination is $p_{15} = 6.6$ W/kg with 50 Hz and 1.5 T. The air gap length is 5 mm. The rated values of this generator are given in Table 5.16.

The stator winding overhang is presented in Fig. 5.32. The constant dimensions are also shown in this figure. The clearance between the coils is 4 mm. The bending angle α is as follows:

$$\alpha = \arcsin \frac{b_4 + 4 \, \text{mm}}{b_4 + b_d}, \tag{5.50}$$

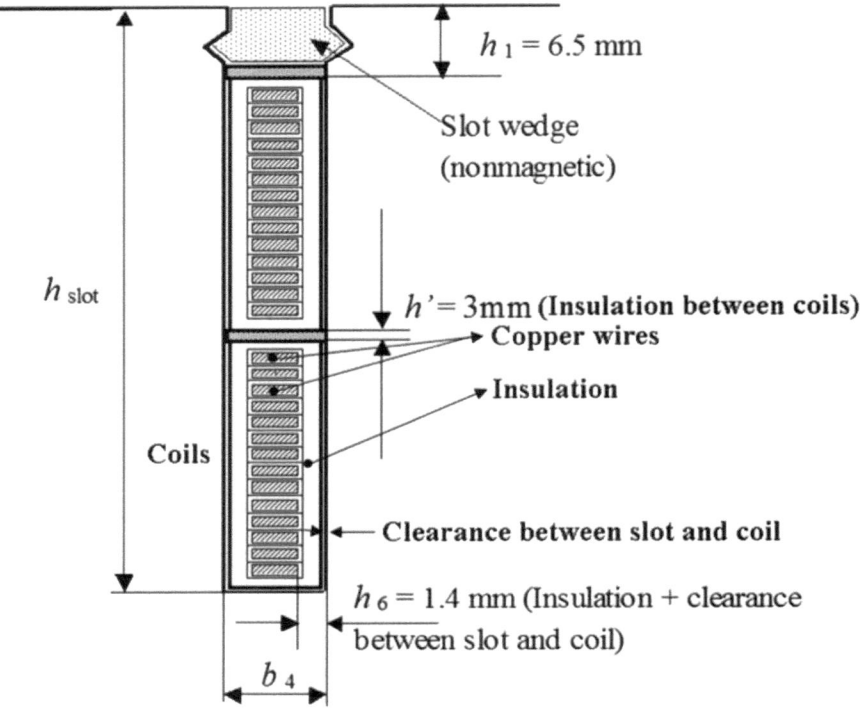

Fig. 5.30 Stator slot form and constant dimensions of slot

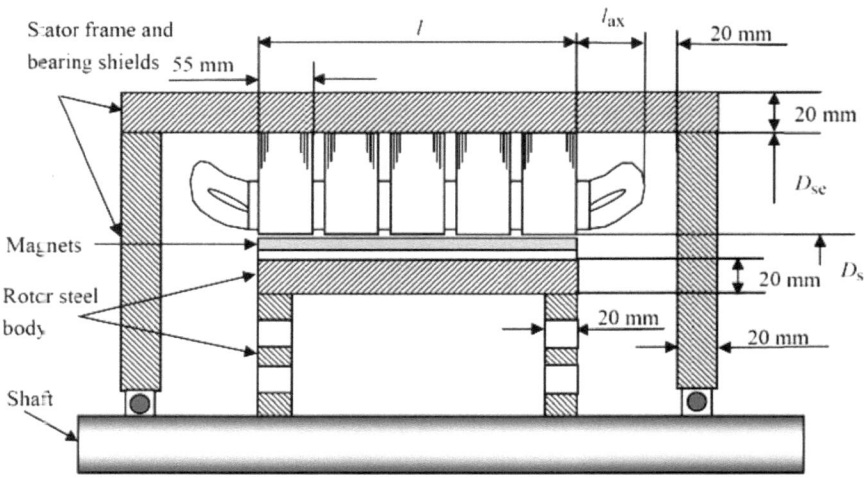

Fig. 5.31 Axial cross section of permanent magnet generator

Table 5.16 Rated values of generator

Power	3 MW
Voltage	690 V
Connection	Star
Speed	16.98 rpm
Number of phases	3

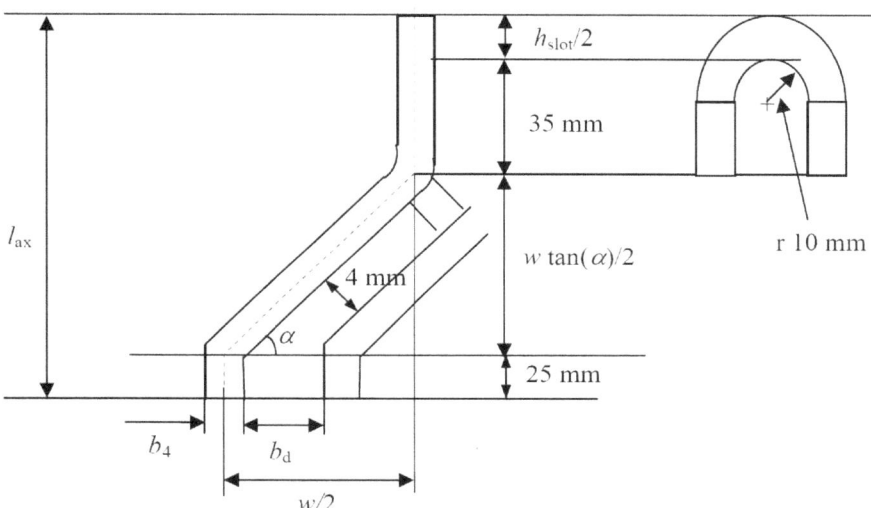

Fig. 5.32 The overhang form and dimensions of the stator winding

where b_4 is the slot width and b_d is the tooth width. The coil span at the middle of the slot is

$$w = \frac{\pi(D_s + h_{\text{slot}})}{2p},\tag{5.51}$$

where D_s is the air gap diameter, h_{slot} is the slot height, and p is the pole pair number. The axial length of the overhang is as follows:

$$l_{\text{ax}} = \frac{w \tan \alpha}{2} + \frac{h_{\text{slot}}}{2} + 60\,\text{mm}.\tag{5.52}$$

The average conductor length of a winding turn is

$$l_{\text{av}} = 2\left(l + \frac{w}{\cos \alpha} + \pi\frac{h_{\text{slot}}}{2} + 131\,\text{mm}\right),\tag{5.53}$$

where l is the length of the stator core (Fig. 5.31).

Table 5.17 Unit prices of materials and capitalized loss costs

Electrical steel (k_{Fe})	4 €/kg
Copper (k_{Cu})	12 €/kg
NdFeB magnets (k_{PM})	60 €/kg
Stator frame and rotor steel body (k_{Fef})	2 €/kg
Losses (k_{Loss})	2 €/W

5.4.4.2 Objective Function and Design Parameters

The design principles of the above wind generator are given in [27]. The objective function $f(x)$ (in €) to be minimized is the sum of the material costs and capitalized costs of the total losses of the generator:

$$f(\mathbf{x}) = k_{Fe}m_{Fe} + k_{Cu}m_{Cu} + k_{PM}m_{PM} + k_{Fef}m_{Frame} + k_{Loss}P_{tot}, \qquad (5.54)$$

where m_{Fe}, m_{Cu}, m_{PM}, and m_{Frame} are the masses of the stator iron core, the stator winding, the permanent magnets, and the stator frame and rotor body, respectively, k_{Fe}, k_{Cu}, k_{PM}, and k_{Fef} are the unit prices of the stator core, the copper, the permanent magnets, and the stator frame and rotor body, respectively, and k_{Loss} is the capitalized costs of the losses (Table 5.17). The cost of the stator core actually includes the punching, the waste parts of the sheet, as well as the assembly of the stator core. The manufacturing cost of the winding is taken into account in the copper cost. The permanent magnet cost includes the corrosion protection, the assembly into bigger cassettes, and the magnetization of the magnets. The stator frame and rotor body costs consist of the material cost and cost of manufacturing the frame and body.

The stator resistive losses are calculated at the temperature of 100 °C, and the iron losses in the stator teeth are as follows:

$$P_{Fed} = 2 \cdot p_{15}(B_d/1.5 \text{ T})^2 (f/50 \text{ Hz})^{1.5} m_d, \qquad (5.55)$$

where p_{15} is the iron loss factor, B_d is the maximum flux density in the teeth, f is the frequency, and m_d is the mass of the stator teeth. The iron losses in the stator yoke are as follows:

$$P_{Fey} = 1.5 \cdot p_{15}(B_y/1.5 \text{ T})^2 (f/50 \text{ Hz})^{1.5} m_y, \qquad (5.56)$$

where B_y is the maximum flux density in the yoke and m_y is the mass of the yoke. The losses in the permanent magnets are assumed to be 1 % of the rated power, i.e., 30 kW. The additional losses are assumed to be 3 % of the rated power, i.e., 90 kW. The friction and ventilation losses are

$$P_\rho = 10 \cdot D_r (l + 0.6 \cdot \tau_p)(\pi n D_r)^2 [\text{W}], \qquad (5.57)$$

Table 5.18 Design parameters with ranges

Parameters	Symbols	Ranges
Stator outer diameter	D_{se}	3.0–15.0 m
Stator core length including the ventilation ducts	l	0.2–3.0 m
Stator yoke height	h_{ys}	0.01–0.5 m
Rotor yoke height	h_{yr}	0.01–0.5 m
Stator slot height	h_{slot}	0.07–0.3 m
Stator slot width	b_4	0.007–0.04 m
Maximum flux density in air gap	B_{max}	0.4–0.9 T
Number of effective conductors in stator slot	z_s	8–26
Number of pole pairs	p	20–100

Table 5.19 Optimization constraints

Stator tooth width	>8 mm
Stator yoke flux density	<2.2 T
Rotor yoke flux density	<2.2 T
Stator tooth flux density	<2.2 T
Maximum output power	>4.8 MW

where D_r is the outer rotor diameter (m), τ_p is the pole pitch (m), and n is the rotational frequency of the rotor (1/s) [28]. Table 5.18 gives the nine design parameters to be optimized and their valid ranges. A total of five practical constraints are also provided in Table 5.19.

5.4.5 HS-CA-Based Optimal Wind Generator Design

In this section, we investigate the effectiveness of the proposed HS-CA in the above-optimal wind generator design problem. The HS coefficients are chosen as the same as in the last section. The CA parameters of Top and β in the HS-CA are 10 % and 0.5, respectively. After a total of 100 independent trials have been run, the average convergence procedures of the HS and four variants of our HS-CA within 1,000 and 10,000 NFE are illustrated in Figs. 5.33 and 5.34, respectively. Figures 5.35 and 5.36 show, respectively, the corresponding optimal costs acquired by the HS and HS-CA during the 100 trials. Note that the costs in these two figures have been ranked. Particularly, Tables 5.20 and 5.21 present the optimal wind generator parameters and costs obtained after 1,000 and 10,000 NFE, respectively. The best costs, worst costs, and average costs obtained in these two cases are also summarized in Tables 5.22 and 5.23.

It is well known that the stator outer diameter, D_{se}, is one of the most important design parameters of our wind generator. Therefore, the relationship between the

Fig. 5.33 Average
convergence procedures of
HS and HS-CA within
1,000 NFE [*Thick line* HS,
solid line HS-CA (N_s), *dotted
line* HS-CA (S_d), *dash-dot
line* HS-CA ($N_s + S_d$),
dashed line HS-CA
($N_s - N_d$)]

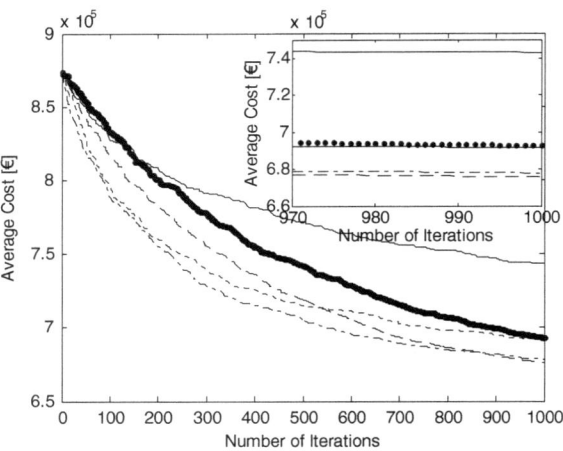

Fig. 5.34 Average
convergence procedures of
HS and HS-CA within
10,000 NFE [*Thick line* HS,
solid line HS-CA (N_s), *dotted
line* HS-CA (S_d), *dash-dot
line* HS-CA ($N_s + S_d$),
dashed line HS-CA
($N_s + N_d$)]

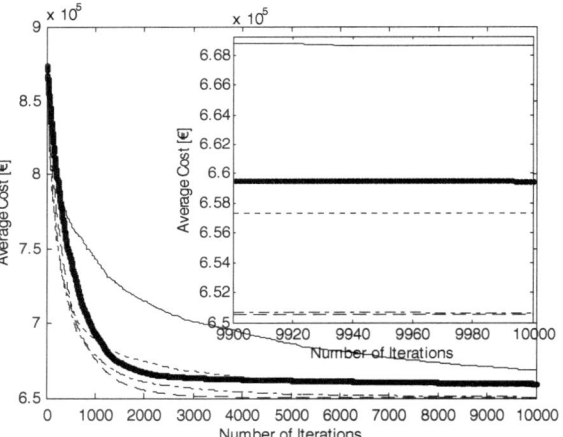

Fig. 5.35 Optimal costs
acquired by HS and HS-CA
within 1,000 NFE [*'Circle'*
HS, *'x-mark'* HS-CA (N_s),
'star' HS-CA (S_d), *'plus'*
HS-CA ($N_s + S_d$), *'square'*
HS-CA ($N_s + N_d$)]

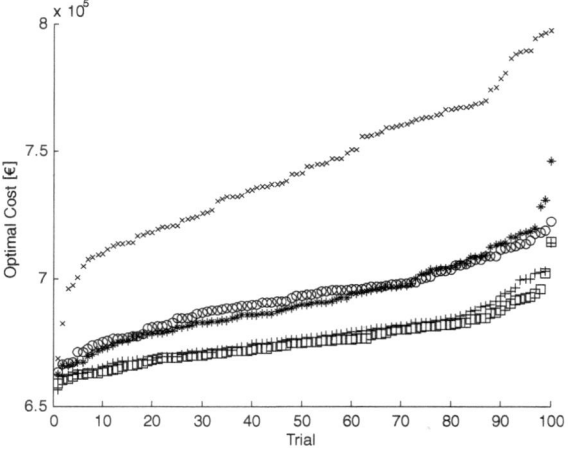

Fig. 5.36 Optimal costs acquired by HS and HS-CA within 10,000 NFE ['*Circle*' HS, '*x-mark*' HS-CA (N_s), '*star*' HS-CA (S_d), 'plus' HS-CA ($N_s + S_d$), 'square' HS-CA ($N_s + N_d$)]

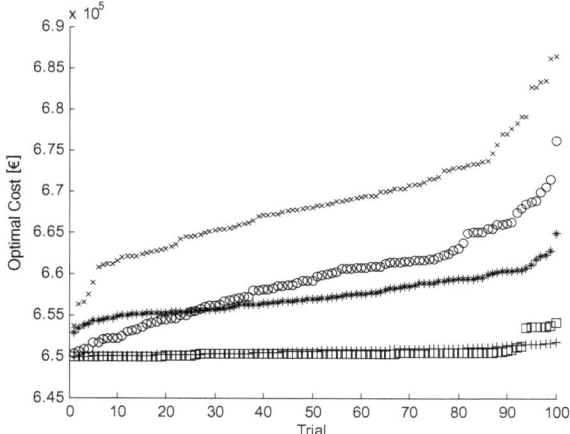

Table 5.20 Optimal parameters and costs acquired by HS and HS-CA within 1,000 NFE

	HS	HS-CA (N_s)	HS-CA (S_d)	HS-CA ($N_s + S_d$)	HS-CA ($N_s + N_d$)
l	0.4548	0.5288	0.5471	0.4687	0.4721
h_{ys}	0.0666	0.0539	0.0343	0.0416	0.0334
D_{se}	8.1120	7.9314	8.4834	8.1119	8.1357
h_{slot}	0.1137	0.0965	0.0991	0.1189	0.0920
B_{max}	0.7057	0.7079	0.6188	0.6914	0.6613
Z_s	22	22	22	20	22
h_{yr}	0.0351	0.0225	0.0248	0.0294	0.0455
p	70	82	95	76	75
b_4	0.0153	0.0102	0.0110	0.0123	0.0144
Costs	6.6350×10^5	6.6869×10^5	6.6243×10^5	6.5630×10^5	6.5917×10^5

Table 5.21 Optimal parameters and costs acquired by HS and HS-CA within 10,000 NFE

	HS	HS-CA (N_s)	HS-CA (S_d)	HS-CA ($N_s + S_d$)	HS-CA ($N_s + N_d$)
l	0.4691	0.4436	0.4569	0.4719	0.4785
h_{ys}	0.0386	0.0455	0.0396	0.0387	0.0380
D_{se}	8.1802	8.6117	8.6409	8.3130	8.2067
h_{slot}	0.1021	0.1034	0.0990	0.1002	0.0999
B_{max}	0.6900	0.6606	0.6722	0.6735	0.6779
Z_s	22	22	22	22	22
h_{yr}	0.0288	0.0306	0.0221	0.0260	0.0260
p	75	72	95	78	77
b_4	0.0134	0.0139	0.0113	0.0133	0.0134
Costs	6.5029×10^5	6.5366×10^5	6.5286×10^5	6.4993×10^5	6.4988×10^5

Table 5.22 Best, worst, and average costs acquired by HS and HS-CA within 1,000 NFE

	HS	HS-CA (N_s)	HS-CA (S_d)	HS-CA ($N_s + S_d$)	HS-CA ($N_s + N_d$)
Best costs	6.6350×10^5	6.6869×10^5	6.6243×10^5	6.5630×10^5	6.5917×10^5
Worst costs	7.2213×10^5	7.9717×10^5	7.4591×10^5	7.1400×10^5	7.1417×10^5
Average costs	6.9266×10^5	7.4301×10^5	6.9166×10^5	6.7784×10^5	6.7574×10^5

Table 5.23 Best, worst, and average costs acquired by HS and HS-CA within 10,000 NFE

	HS	HS-CA (N_s)	HS-CA (S_d)	HS-CA ($N_s + S_d$)	HS-CA ($N_s + N_d$)
Best cost	6.5029×10^5	6.5366×10^5	6.5286×10^5	6.4993×10^5	6.4988×10^5
Worst cost	6.7621×10^5	6.8656×10^5	6.6497×10^5	6.5182×10^5	6.5418×10^5
Average costs	6.5947×10^5	6.6862×10^5	6.5733×10^5	6.5065×10^5	6.5052×10^5

Fig. 5.37 Relationship between optimal costs and stator outer diameter (D_{se})

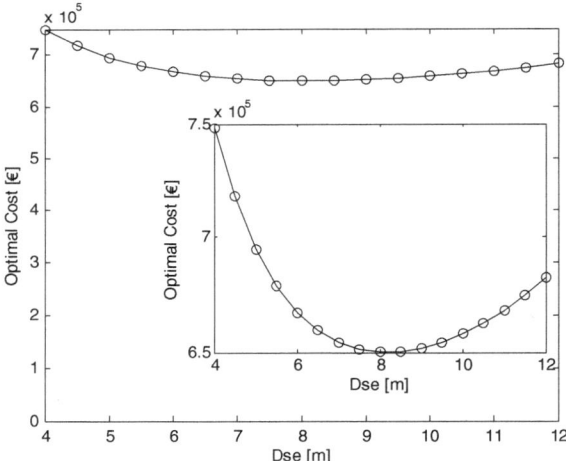

optimal costs and D_{se} is demonstrated in Fig. 5.37. For each D_{se} in Fig. 5.37, the optimal cost is chosen from 100 independent 10,000 NFE trials on the basis of HS-CA ($N_s + N_d$). We can observe that at the beginning, with D_{se} growing from 4 m the cost acquired becomes smaller and smaller. The best one can be obtained, when D_{se} is approximately 8.2 m. Nevertheless, this relationship curve is rather flat. If the optimal outer diameter 8.2 m is halved to 4 m, the capitalized cost will increase by only around 15 %.

As we know, the optimization performance of the CA heavily depends on its parameters, such as β. Different CA coefficient values applied can result in

Fig. 5.38 Relationship between optimal costs and parameter β in HS-CA $(N_s + N_d)$

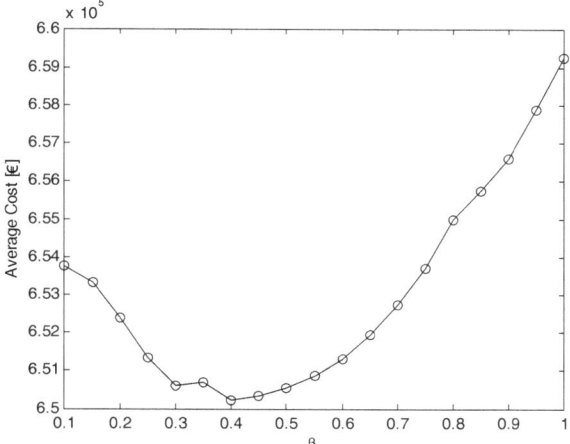

significant differences in the optimal solutions ultimately acquired. Figure 5.38 shows the relationship between the optimal costs obtained by the HS-CA $(N_s + N_d)$ and varying β, when β grows from 0.1 to 1. Unfortunately, β is usually an application-dependent parameter, and there is no analytic way yet to choose its best value for a specific problem. Thus, it is generally determined based on a *trial-and-error* procedure, which might be time-consuming in practice.

Apparently, compared with the original HS method, the three variants of our HS-CA (HS-CA (S_d), HS-CA $(N_s + S_d)$, and HS-CA $(N_s + S_d)$ can achieve moderately better average optimization results within the same NFE, because of the CA-based utilization of the search knowledge extracted from the HS. For example, HS-CA $(N_s + S_d)$ offers improvements of about 2.4 and 1.4 % in the average-optimized cost in Tables 5.16 and 5.17, respectively. Nevertheless, the HS-CA has a higher computation complexity than that of the HS algorithm, due to the incorporated CA operations. Furthermore, as aforementioned, the performance of the HS-CA is indeed affected by the *Influence* function used, which plays a pivotal role in its optimization capability. As a matter of fact, from Tables 5.22 and 5.23, we can find out that HS-CA (N_s) is even worse than the original HS method. To summarize, in this optimal design of wind generator problem, the optimization capabilities of the four HS-CA variants can be ranked as follows: HS-CA $(N_s + N_d)$ > HS-CA $(N_s + N_d)$ > HS-CA (S_d) > HS-CA (N_s).

References

1. X.Z. Gao, X. Wang, H. Xu et al., Multi-modal optimization using a modified harmony search method. J. Inf. Comput. Sci. **10**(6), 1651–1664 (2013)
2. S.W. Mahfoud, A Comparison of Parallel and Sequential Niching Methods. in *The Sixth International Conference on Genetic Algorithms*, San Francisco, CA, 15–19 June 1995

3. S.W. Mahfoud, Niching Methods for Genetic Algorithms. Dissertation, University of Illinois at Urbana-Champaign, 1995
4. Z.W. Geem (ed.), *Recent Advances in Harmony Search Algorithm* (Springer, Heidelberg, 2010)
5. X.Z. Gao, X. Wang, S.J. Ovaska et al., A hybrid optimization method of harmony search and opposition-based learning. Eng. Optim. **44**(8), 895–914 (2012)
6. X.Z. Gao, X. Wang, T. Jokinen et al., A hybrid optimization method for wind generator design. Int. J. Innov. Comput. Inf. Control **8**(6), 4347 (2012)
7. X.Z. Gao, X. Wang, S.J. Ovaska, Uni-modal and multi-modal optimization using modified harmony search methods. Int. J. Innov. Comput. Inf. Control **5**(10a), 2985–2996 (2009)
8. Z. Michalewicz, *Genetic Algorithms + Data Structures = Evolution Programs*, 3rd edn. (Springer, Berlin, 1996)
9. K.N. Krishnanand, D. Ghose, Glowworm swarm optimization for simultaneous capture of multiple local optima of multimodal function. Swarm. Intel. **3**(2), 87–124 (2009)
10. H.R. Tizhoosh, Opposition-based learning: a new scheme for machine intelligence. in *International Conference on Computational Intelligence for Modelling Control and Automation*, Vienna, Austria, Nov 2005
11. H.R. Tizhoosh, Opposition-based reinforcement learning. J. Adv. Comput. Intel. Intel. Inf. **10**(5), 578–585 (2006)
12. S. Rahnamayn, H.R. Tizhoosh, M. Salama, A novel population initialization method for accelerating evolutionary algorithms. Comput. Math. Appl. **53**(10), 1605–1614 (2007)
13. S. Rahnamayan, H.R. Tizhoosh, M.A. Salama, Opposition-based differential evolution. IEEE Trans. Evol. Comput. **12**(1), 64–79 (2008)
14. M.M. Ali, C. Khompatraporn, Z.B. Zabinsky, A numerical evaluation of several stochastic algorithms on selected continuous global optimization test problems. J. Glob. Optim. **31**(4), 635–672 (2005)
15. X.Z. Gao, X. Wang, S.J. Ovaska et al., A modified harmony search method in constrained optimization. Int. J. Innov. Comput. Inf. Control **6**(9), 4235–4247 (2010)
16. X. Wang, X.Z. Gao, S.J. Ovaska, Fusion of clonal selection algorithm and harmony search method in optimization of fuzzy classification systems. Int. J. Bio-Inspired. Comput. **1**(1–2), 80–88 (2009)
17. R.A. Fisher, The use of multiple measurements in taxonomic problem. Ann. Eugen. **7**, 179–188 (1936)
18. X. Chang, J.H. Lilly, Evolutionary design of a fuzzy classifier from data. IEEE Trans. Syst. Man. Cybern. Part B Cybern. **34**(4), 1894–1906 (2004)
19. R.G. Reynolds, B. Peng, Cultural algorithms: modeling how cultures learn to solve problems. in *IEEE International Conference on Tools with Artificial Intelligence*, Boca Raton, FL, 15–17 Nov 2004
20. R.G. Reynolds, C.J. Chung, CAEP: an evolution-based tool for real-valued function optimization using cultural algorithms. Int. J. Artif. Intel. Tools **7**(3), 239–293 (1998)
21. R.G. Reynolds, C.J. Chung, Knowledge-based self-adaptation in evolutionary programming using cultural algorithms. in *IEEE International Conference on Evolutionary Computation*, Indianapolis, IN, 13–16 Apr 1997
22. D.E. Goldberg, *Genetic Algorithms in Search, Optimization, and Machine Learning, Reading* (Addison-Wesley, Boston, 1989)
23. O. Kramer, Evolutionary self-adaptation: a survey of operators and strategy parameters. Evol. Intel. **3**(2), 51–65 (2010)
24. A. Hedar, Kyoto University (2005). Retrieved from http://www.optima.amp.i.kyotou.ac.jp/member/student/hedar/Hedar_files/TestGO.htm
25. K. Deb, Optimal design of a welded beam via genetic algorithms. J. Am. Inst. Aeronaut. Astronaut. **29**(11), 2013–2015 (1991)
26. C.A.C. Coello, Constraint-handling using an evolutionary multiobjective optimization technique. Civ. Eng. Environ. Syst. **17**(4), 319–346 (2000)

27. J. Pyrhönen, T. Jokinen, V. Hrabovcová, *Design of Rotating Electrical Machines* (Wiley, West Sussex, 2008)
28. X.Z. Gao, T. Jokinen, X. Wang et al., A new harmony search method in optimal wind generator design. in *XIX International Conference on Electrical Machines*, Rome, Italy, 6–8 Sept 2010
29. M. Setnes, H. Roubos, GA-fuzzy modeling and classification: complexity and performance. IEEE Trans. Fuzzy Syst. **8**(5), 509–522 (2000)
30. Y. Shi, R. Eberhart, Y. Chen, Implementation of evolutionary fuzzy system. IEEE Trans. Fuzzy Syst. **7**(2), 109–119 (1999)
31. M. Russo, Genetic fuzzy learning. IEEE Trans. Evol. Comput. **4**(3), 259–273 (2000)

Chapter 6
Conclusions

Abstract In the harmony search (HS) method, there are only three distinguishing factors: harmony memory considering rate (HMCR), pitch adjusting rate (PAR), and HS memory (HM), used to control the quality of solutions and balance the intensification and diversification during the search process. These three operators determine the attractive properties of the HS including structure simplicity and fast convergence, which also promote the research of improved HS algorithms and fusion strategies with other nature-inspired computational (NIC) approaches. However, the consideration of balance intensification and diversification is inevitably a challenging problem in the hybrid NIC algorithms that possesses the advantages of robustness, efficiency, and accuracy, and these objectives usually conflict each other. Additionally, the 'no free lunch (NFL)' theorem should also be taken into account when handling various optimization problems in engineering. This chapter gives a concise summary of the book and points out the further research trend of the HS method.

Keywords Harmony search method · Hybrid harmony search methods · No free lunch · Optimization applications

6.1 Summary of This Book

Inspired by certain natural phenomena and biological models, the NIC methods have become the dominant approaches over the conventional numerical algorithms, due to their characteristics, e.g., flexibility, randomness, robustness, intuitive guidelines, and derivative-free. The simple computation and structure of the HS have promoted it to be a new emerging research focus of the NIC methods in the recent years. This book has presented a brief introduction to the HS algorithm as well as its variations in different optimization applications.

In original HS method, there are only three distinguishing elements: HMCR, PAR, and HM, used to control the quality of solutions and further properly balance the intensification and diversification during the search process. However, its

© The Author(s) 2015   85
X. Wang et al., *An Introduction to Harmony Search Optimization Method*,
SpringerBriefs in Computational Intelligence,
DOI 10.1007/978-3-319-08356-8_6

inherent drawbacks (slow convergence, outdated memory information, etc.) motivate the research in the improved HS methods. The HS variations can be based on motivations of these three operators. In this book, four typical hybrid HS methods fused with niching technique, opposition-based learning, clonal selection algorithm, and cultural algorithm are introduced and explored. These hybrid approaches have been demonstrated to achieve significant improvement in the performance compared with the original HS in the applications of multimodal, high-dimensional benchmark function, minimization of weight of a tension/compression spring, optimization of Sugeno fuzzy classification systems, optimal design of welded beam, gear train, pressure vessel, and permanent magnet direct-driven wind generator. Therefore, they are supposed to have promising potentials in coping with a large variety of engineering optimization problems.

6.2 Further Research

It has been demonstrated that the HS after the modification and hybridization is well capable of dealing with various problems in different optimization fields. According to the current literature, the collaboration with other NIC methods dominates in the development of the HS, which is usually in the manner of cooperator or embedded operators. The cooperator in the hybrid HS systems promotes the information sharing and exchange among different algorithms so as to achieve an improved searching efficiency. However, the hierarchical configuration containing the embedded operator emphasizes the task of global and local search. This fusion strategy has allowed us to overcome the drawbacks of stand-alone HS algorithm without sacrificing its advantages, and it helps to create a more compact and reconfigurable structure and yields the enhanced convergence performance. On the other hand, the common problem existing in all the meta-heuristic algorithms, i.e., mutually balancing intensification and diversification in the solution space, is still unavoidable in the hybrid HS methods developed and studied so far.

It is well known that the 'no free lunch (NFL)' theorem is a fundamental barrier to the exaggerated claims of the power and efficiency of any specific optimization algorithm. In other words, there is no single optimization method in practice that can be the best for all kinds of engineering problems, because whatever an algorithm gains in performance in one class of problems is necessarily offset by poor performance in the remaining ones. One possible way to handle the negative implication of this NFL theorem is to restrict the applications of a given algorithm including the HS and its variants to only a particular type of tasks. In addition, an appropriate trade-off between the intensification and diversification should be made in examining the hybrid HS.

Index

A
Algorithms, 1–3, 7–9, 13, 17, 19, 22, 24, 34, 50, 53, 59, 60, 81, 85, 86
Ant colony optimization (ACO), 23, 24

B
Bandwidth (bw), 10
Belief space, 59, 60, 61, 63
Benchmark functions, 8, 31
Benchmarks, 8, 41, 64

C
Clonal selection algorithm (CSA), 15–20, 50, 51, 53, 56, 57, 59
Clonal selection principle (CSP), 15, 16
Constrained optimization, 24, 25, 28
Convergence, 8, 13, 14, 17–20, 38, 41, 44, 50, 53, 59, 63, 64, 68, 77, 86
Crossover, 7, 9, 13, 14, 31, 32, 60
Cultural algorithm (CA), 59–62, 64, 68, 69, 77, 80

D
Deterministic crowding (DC), 31–34, 37
Differential evolution (DE), 8, 24, 39
Diversification, 3, 85
Diversity, 14, 17, 20–22, 32, 34

F
Feasible solutions, 25
Fisher iris data, 50, 52, 53, 55, 56
Fitness, 8, 4, 17, 21, 25, 31, 32, 39, 41, 50, 61–63
Fuzzy classification, 50–53, 55, 57, 59, 86

G
Gear train design, 70, 71
Genetic algorithm (GA), 7–9, 14, 17, 19

H
Harmony memory (HM), 6–8, 21, 23, 25, 26, 28, 40, 41, 50, 61
Harmony memory considering rate (HMCR), 6, 7, 23, 24, 34, 37, 44, 64
Harmony search (HS) method, 3, 6–9, 22, 24, 25, 32, 33, 37, 41, 50, 59, 68, 85, 86
Hybrid harmony search (HS) method, 8, 38–40, 50–54, 57, 59, 86
Hybrid nature-inspired computational (NIC) methods, 1–3, 20, 23, 85
Hybridization, 3, 8, 20, 23, 24, 38, 40, 59, 61, 86

I
Infeasible solutions, 26, 27
Influence function, 62–64, 68, 69, 81
Intensification, 3, 85, 86

K
Knowledge, 23, 32, 59–63

M
Meta-heuristic methods, 2, 5
Multimodal optimization, 31–34, 37
Mutation, 7, 9, 14, 16–18, 24, 40, 59, 60, 62, 63

N
Nature-inspired computational (NIC) algorithms, 1–3, 20, 23, 85

© The Author(s) 2015
X. Wang et al., *An Introduction to Harmony Search Optimization Method*,
SpringerBriefs in Computational Intelligence,
DOI 10.1007/978-3-319-08356-8